CONTENTS

Introduction to the guide 2

NEWS
News 3
Ultra HD/4K and High Dynamic Range TV 14

DIGITAL TV – THE OPTIONS
Digital platform guide 16
Freeview 18
Freeview Play 20
Humax FVP-4000T review 21
YouView 22
Freesat 24
Sky 26
Sky's new Q product range 28
Recording on Sky 29
Virgin Media 30
BT TV 32
TalkTalk TV 33
EE TV 34

INTERNET TV
Amazon Fire TV 35
Internet television 36
Apple TV 38
NOW TV 39
Chromecast 40
Google Nexus Player 41
BBC iPlayer 42

CHANNEL GUIDES
Freeview channels 44
Freesat channels 46
Sky channels 48

TV EQUIPMENT GUIDES
Buying a new television 52
Television types 56

.... 62
Video discs 64
Home cinema and TV sound 66

HOW TO
Connecting your TV equipment 70
Cables and connectors 74
TV calibration 76
Sending video signals around the house 78
TV and audio troubleshooting 80
Interference and picture problems 82
Archiving recordings made on video tape 87
Help with sight problems 88
Help with hearing problems 90

AERIAL AND SATELLITE DISH INFO
Buying a new aerial 92
Setting up a new aerial 96
TV reception in shared properties 98
Television on the move 100
How to set up a satellite dish 106
Sky's installer set-up menu 111

TRANSMITTER INFORMATION
Freeview transmitter map 112
Freeview transmitter sites 113
Freeview local TV multiplexes 142
Local television stations 144
Saorview, Irish digital TV 146

CONTACT ADDRESSES AND GLOSSARY
Contacts and further information 148
Glossary / abbreviations 156

ORDER FORM
Order form 160

INTRODUCTION TO THE GUIDE

Welcome to the Television Viewer's Guide. You'll find the latest information about television, along with equipment advice and guides to help if you are looking to purchase a new television, set-top box, Blu-ray player or digital recorder.

This year's must-have technologies, being pushed by the industry, are Ultra HD/4K, and HDR – you'll find information about both in the guide. To future-proof new purchases so that they work with the new technology, make sure that they have a suitable HDMI port – v2.0a at the minimum.

General TV buying guide
You'll find a guide covering the key things you should consider if you are looking for a new TV set. It starts on page 52. You'll find more detailed information about new TV technology elsewhere in the guide, but this section is a good starting point.

Which platforms are best?
The main TV platforms, Freeview, Sky, Freesat and Virgin are still the most popular way to pick up broadcast TV. You'll find a simple outline showing the key features, and pros and cons of each platform.

October saw an interesting development – the launch of Freeview Play. It offers a good way to watch the main broadcast channels in a package that also includes recording and the best catch-up TV platforms, such as BBC iPlayer. See page 20 for further information.

'Smart' services
Internet delivered 'Smart' services continue to grow and are often incorporated into packages from the main TV providers. The amount of content available is increasing too. This includes programming in Ultra HD.

Many of the main internet players have launched new and improved versions of their devices this year. Notable are the new Google Chromecast, Amazon Fire TV, NOW TV, and Apple TV devices. Details are in the guide.

If want to dip your toes into the smart TV world, we'd suggest that you start off by finding a platform that offers BBC iPlayer - it's our favourite internet TV service by far.

If you are watching TV online consider your broadband costs and speed. If you've an unlimited broadband allowance there's no problem. If not it's worth upgrading. Likewise it's no fun watching catch-up TV on a slow connection that's forever stopping and starting – look for fibre or cable for the fastest speeds.

Jargon buster
Throughout the guide we have aimed to keep our explanations as simple and as clear as possible. You'll find the most commonly used abbreviations in the Glossary on page 156.

As usual if you have any suggestions to improve the guide, or have information you think should be added, please let us know.

Clive Woodyear   Editor

Television Viewer's Guide
PO Box 888, Plymouth PL8 1YJ
Telephone: 01752 872888
email sales@radioguide.co.uk
website www.radioguide.co.uk
Editor Clive Woodyear
Technical Editor Eugene Trundle
ISBN 978-1-871611-79-3
Printed by Warners, The Maltings, Manor Lane, Bourne, Lincolnshire PE10 9PH
Cover Design Chris Judson
Published by, and © Clive Woodyear 2015. No part of this publication may be reproduced in any form, without prior written permission from the publisher.
Disclaimer Whilst every effort is made to ensure that the information in this magazine is accurate, the editor/publisher or their agents cannot accept any liability for loss or damage caused by errors in, or from, the information given.
Order form You will find an order form on page 160. You can also order copies of the guide via the website.
Acknowledgements We could not have produced this guide without help from many individuals and organizations; including Mark Carver, Kevin Ryan, David Sullivan, Dan Williams, Bill Wright, Phil Wright and Gill Woodyear. Thanks also to BBC Reception Advice, The Digital Television Group, Ofcom and manufacturers who have lent equipment.

NEWS

GENERAL NEWS

EU probes TV makers over energy efficiency test scores

In October the European Commission announced that it was following up two reports that raise concerns that software used in TVs may be skewing their energy rating scores.

One study indicates that some Samsung TVs nearly halve their power consumption when a standardised test is carried out. Another accuses a different, unnamed manufacturer, of adjusting the brightness of its sets when they recognise the test film involved.

Samsung has denied any wrongdoing. It acknowledged that it used software that altered its televisions' performance during tests, but said this was the effect of a general energy efficiency feature that came into effect during normal use and had nothing to do with the testing process.

However, one environmental campaign group has likened the accusations to the Volkswagen diesel scandal, in which the German car firm admitted to programming its cars to deliberately cheat emissions tests.

More signed programmes for deaf television viewers

A change in Ofcom rules will boost the amount of TV programmes made specifically for sign language users. The decision will provide substantial benefits to deaf and hard-of-hearing TV viewers in the UK.

Currently smaller TV channels – those with an audience viewing share of between 0.05% and 1% – must show 30 minutes of programmes each month presented in British Sign Language (BSL). As an alternative they can contribute a minimum of £20,000 a year to help make sign-presented TV programmes available.

Most choose to support the British Sign Language Broadcasting Trust (BSLBT), which funds sign-presented content shown on the Community Channel and Film 4. This content can also be viewed on the BSLBT website.

From January 2016 the amount of sign-presented programmes on smaller channels will begin to increase from 30 to 75 minutes per month.

Voluntary licence fee for over-75s

The BBC says it will only face a 10% cut in licence fee funding for its services, once other aspects of the licence fee deal are taken into account – less than the amount claimed across the media in the wake of the Summer Budget.

With the BBC budget under pressure and the corporation taking over the funding for TV licences for the over-75s, those entitled to a free TV licence will be given an opportunity to voluntarily pay for a licence, according to James Heath, the BBC's Director of Policy.

Samsung warning over 'listening' TV

In February this year Samsung warned customers about discussing personal information in front of their smart television set! The warning applies to TV viewers who control their Samsung Smart TV using its voice activation feature. When the feature is active, such TV sets 'listen' to what is said and may share what they hear with Samsung or third parties.

On 10 February it felt compelled to update its privacy policy after concerns that its TVs were recording and transmitting everything said in front of them.

The warning explained that the TV set will be listening to people in the same room to try to spot when commands or queries are issued via the remote. It goes on to say: 'If your spoken words include personal or other sensitive information, that information will be among the data captured.'

Samsung is not the first maker of a smart, net-connected TV to run into problems with the data the set collects. In late 2013, a UK IT consultant found his LG TV was gathering information about his viewing habits. Publicity about the issue led LG to create a software update which ensured data collection was turned off for those who did not want to share information.

BBC testing mind control TV

The BBC has been experimenting to see if people can control televisions using nothing but their brainwaves.

BBC Digital has been working with a company known as The Place, with the aim of giving programme makers, technologists and other users an idea of how the technology might be used in the future.

The first trial run saw 10 BBC staff members try out the app, and all were able to launch BBC iPlayer and start viewing a programme simply by using their minds. It was much easier for some than it was for others, but they all managed to get it to work.

The electroencephalography (EEG) brainwave reading headset worn by the user has a small sensor that measures electrical activity in the brain.

One of the objectives is to help users with a broad range of disabilities who cannot easily use traditional TV remote controls or other conventional interfaces. In this instance there may be help for people affected by motor-neurone disease or suffering locked-in-syndrome.

Previously the BBC has experimented with voice control with BBC iPlayer on Xbox One.

TELEVISION TECHNOLOGY

What is HDR and when will it get onto our screens?

HDR, or High Dynamic Range, isn't a competitor to 4K Ultra HD, but rather an additional technology that will become available on some 4K TVs that could improve the pictures you see on screen.

The term HDR is already used in the world of photography. An HDR photo usually combines several images taken at different exposures into a single picture that appears to have even more detail and sharpness.

The system works similarly with video in compatible HDR TVs, creating a greater dynamic range between the blackest blacks and the whitest whites, plus all the subtle tones in between. This means that TV pictures will seem even more detailed, particularly on darker scenes.

There are likely to be problems with the bandwidth demands for HDR content delivered across the internet. Netflix, one of the key suppliers of 4K content is suggesting that there will be an additional 20% bandwidth requirement for HDR on top of Ultra HD content, which could push the requirements up to around 30Mbps.

At the moment HDR is struggling to get onto our TVs as Amazon Prime is the only service offering accessible content – and even then only a handful of shows. But the standards have yet to be finalised and when they are Netflix and other platforms are expected to provide content too.

Commercial channels concerned by EE TV's Replay

EE TV's Replay feature – which continually records 24 hours of your favourite channels – is said to be delaying negotiations over catch-up players on the service.

ITV, Channel 4 and Channel 5 are reported to be concerned that Replay lets viewers fast-forward through adverts on hundreds of hours of TV shows instead of watching the ads inserted into their catch-up services.

Final stages of satellite transfer

Numerous TV and radio services broadcasting to the UK and Ireland have changed satellites this year as part of the final stages of a programme to renew the fleet of satellites serving Sky and Freesat users.

The UK and Ireland is served by a fleet of satellites orbiting the earth at 28 degrees East. The changes involved services being removed from the aging Eutelsat 28A satellite, which has recently been broadcasting services at significantly lower power levels than its neighbouring orbiting counterparts. The new satellite Astra 2G arrived at 28 degrees East in mid-June 2015, triggering more changes.

Most viewers in the UK and Ireland using either Freesat or Sky should not have experienced any problems, with the satellite platform operators updating their Electronic Programme Guides (EPGs) to redirect receivers to any new frequencies that may apply.

Previous satellite transfers have resulted in some of the UK's main free-to-air television channels becoming less widely available in continental Europe, following requests from broadcasters to reduce their satellite footprint due to programme rights issues. The changes annoyed numerous expats in southern France, northern Italy and Spain, who were among the worst hit by the last satellite transfers.

All future Freeview products will be HD beyond 2016

In September Freeview announced a migration of its product labelling towards HD-only products from the start of 2016.

NEWS   5

The UK's biggest TV platform said that the Freeview trade mark licence will only be granted for HD products from the end of 2016, and will no longer appear on new SD products.

Samsung and LG using new Quantum dot technology in their TV sets

Samsung has begun to apply quantum-dot technology to its screens. The new technology uses crystals more than 50,000 times smaller than the width of a human hair that can generate vivid colours and are capable of being more than twice as bright as conventional TV screens.

To complicate matters, LG also has new quantum-dot-based LCD TVs of its own, but insists its OLED ones are better. OLED sets have yet to gain a large market share due to the high price of the technology.

The rise and fall of 3D TV

In May Sky announced it would cut its 3D channel from its TV offering, so signalling another nail in the coffin for 3D at home.

The BBC has already shelved its short-lived 3D sports pilot and 3D Blu-ray sales have not been encouraging.

Sky has taken the decision to shelve the channel and move its 3D content to its Sky On Demand service. 3D fans should at least still have some 3D TV available to watch, but the format no longer seems to be a priority for the pay TV provider.

Part of that is a comfort thing – do we really want to wear 3D goggles while sat on the sofa? Despite the fall from grace of 3D TV, the major television manufacturers are still incorporating the feature into this year's flagship sets. Where previously its inclusion was shouted from the rooftops though, it's now hardly mentioned with the majority of the hype going to other new features.

4K Blu-ray players

In July 2015 the Blu-ray Disc Association (BDA) announced the completed specification for Ultra HD Blu-ray.

The new Ultra HD Blu-ray standard will enable the high capacity discs to deliver content at the 4K resolution of 3840 x 2160, with an upgraded colour range. New 4K Blu-ray discs will enable high-dynamic range (HDR) content as well as offering high-frame rate (HFR) playback.

The BDA has mandated all the Ultra HD Blu-ray players will have to be backwards compatible with the current Blu-ray standard. So, you're not going to have to replace your entire library to get access to 4K Blu-ray.

BT Ultra HD TV service only for the fastest fibre connections

BT's new Ultra HD TV 4K channel launched on August 2nd 2015 with the FA Community Shield match between Chelsea and Arsenal, using a Humax 1TB Ultra HD YouView+ set top box for £15 a month.

It will only be available to customers on its Infinity 2 product, with broadband line speeds of no less than 44Mbps.

BT will charge £15 for a package that includes its new Ultra HD channel.

BT is requiring customers to purchase its BT Infinity 1 or 2 broadband package through a line deemed fast enough line to support BT Sport Ultra HD. A £44 intallation charge is required for the 1TB Ultra HD Youview receiver.

However, viewers weren't be able to access Netflix and its UHD offering on the Ultra HD YouView box at launch. BT says that it is working on making it available soon.

First Ultra HD transmission at High Dynamic Range

In May SES and Samsung demonstrated the world's first broadcast of high dynamic range Ultra HD (UHD) TV that uses High Dynamic Range (HDR) technology.

The transmission from an SES satellite at 19.2 degrees East used BBC Research & Development's High Dynamic Range (HDR) technology to deliver the UHD and HDR content directly to a Samsung TV.

Sky prepares European Ultra HD push and DVB-S switch-off

Sky Germany has acquired extra satellite capacity as it prepares to launch an Ultra HD service. The news comes as Sky announces it's going to ditch broadcasts in the DVB-S standard in parts of Europe.

But while Sky UK has previously been the centre of satellite TV innovation, it's Sky Germany that is preparing to get an Ultra HD service on the market.

So far, Sky UK has remained tight-lipped over its eagerly anticipated Ultra HD launch plans for UK and Irish subscribers.

Sharp's Super HD TV

Sharp has announced plans to sell an 8K television screen from October 2015. Although several companies have developed 'super hi-vision' resolution test models, this is the first such TV to be made commercially available.

The 8K format provides 16 times as many pixels as 1080p high definition. It creates an image so detailed that it can appear three-dimensional.

However, the 85in (2.16m) device's £86,000 price is likely to limit sales. Interest is expected to come mainly from broadcasters and other companies involved in testing the format.

NHK pushes 8K trials ahead of Tokyo 2020

Japanese public broadcaster NHK will conduct 8K Ultra HD satellite broadcasting trials from the Rio Olympics and the SuperBowl in 2016 as it prepares for the 2020 Tokyo Olympics.

Market analysts suggest that panels at 7680 x 4320 pixels will be selling by 2019, with 65-inch panels taking most of the 8K UHD market.

CHANNELS

BBC has no plans to close BBC4, says strategy chief

With the BBC looking to make £650 million in savings, the BBC's strategy chief James Purnell said there are no plans to close BBC4 as part of the cuts faced by the corporation but added: 'We are not ruling anything in or out.' Rolling news, BBC4 and the corporation's children's TV channels have been identified as possible candidates for the chop.

The station has seen its budget slashed in recent years and it no longer has its own controller, and is overseen by BBC2 boss Kim Shillinglaw.

The station still has a loyal following for its arts, culture and documentary output, and has looked to innovate with programmes such as its 'Slow TV' season as it attempts to differentiate itself from BBC2.

BBC3 to go online

BBC3 will go online only in the new year as part of a round of budget cuts. There has been widespread opposition from the BBC Three audience and some of the channel's talent, with a petition against the move reaching over 300,000 names.

BBC4's supporters will be hoping it does not suffer the same fate.

BBC4 cost £63m in 2014/15, of which £46m was spent on content, and was watched by just under 14% of the population at least once a week – compared to 47% for BBC2 and 73% for BBC1.

Channel 4 relaunches 4oD as All 4, which includes live mobile streaming

In March 2015 Channel 4 changed the name of its online and on-demand service from 4oD to All 4. All 4 adds a lineup of new features, including live streaming to mobiles and event-based content.

The service has been split into three different sections. 'On demand' is where you'll be able to watch the shows that Channel 4 offers from its channels, as well as archive shows like The IT Crowd and others.

'Now' is where you can stream TV live as well as interact with things happening in real-time. Predictably, 'On soon' is the place to see trailers for upcoming shows, and any shows Channel 4 airs first online.

ITV to launch new online Hub

ITV is to bring all of its online activities under a single brand, the ITV Hub, from later this year. The Hub will replace ITV.com and ITV Player across PC, mobile and connected TV services.

Live streaming of ITV's channels will form a central part of the new service.

New Cambridge TV channel launches

Cambridge's new local TV channel launched in August 2015, the first of the second phase of local TV roll-out in the UK.

Viewers with Freeview, YouView or Virgin Media in the Cambridge area were able to watch the launch of Cambridge TV, billed as the 'new local TV channel for bright, curious people'. It is available on Freeview and YouView channel 8 and Virgin Media channel 159. It is due to be added to Freeview during the day on Tuesday, so that viewers will be able to retune Freeview or YouView in the evening. Virgin Media viewers in Cambridgeshire will see the channel automatically added to their channel line-up.

Viewers outside of the terrestrial and cable coverage area can access Cambridge TV online at www.cambridge-tv.co.uk

Cambridge TV was the 19th local TV channel to launch in the UK since November 2013 and the first of the second phase of local TV channels to go live. Ofcom began the first phase of its licensing process for local TV in 2012, which led to the first local TV stations launching on Freeview in late 2013 / early 2014. Cambridge TV, under the name Cambridge Presents, was announced as the winner of a local TV licence in October 2013, when Ofcom started the second phase of local TV licence awards.

YourTV launches on Freeview, Sky and Freesat

YourTV is a free-to-air entertainment channel from Fox Life, launched on the 1st of October. Its Freeview signal will reach just over 90% of UK households. It is available on Freeview (channel 73), Sky (channel 275), YouView (channel 73) and Freesat (channel 161).

FREEVIEW

Freeview adopts new look

In June digital terrestrial television platform Freeview adopted new branding, including a new Freeview logo that was unveiled earlier this year. This now appears on its web and social media platforms and the new logo will soon begin to appear on the packaging of Freeview-compatible TVs, PVRs and set-top-boxes.

The new look has been adopted ahead of the launch of the new connected TV platform Freeview Play.

Freeview Play launches

Freeview's next-generation smart TV platform, Freview Play, launched in October. It combines catch-up TV from the BBC, ITV, Channel 4 and Channel 5, on-demand services and live television in a subscription-free platform which works on all existing broadband services.

Freeview Play outwardly looks very similar in what it offers to YouView, but it does not have the pay channels associated with the YouView packages available from BT, TalkTalk and Plusnet– and you won't be tied in to a provider in order to keep the service.

Panasonic 2015 Viera TVs were updated automatically to prepare for the launch, and viewers will be able to scroll back through the TV guide for programmes on catch-up or access on-demand programming through the apps page.

Guy North, Managing Director of Freeview, said: We believe that Freeview Play is set to become the new normal way of watching TV. Viewers will be free to choose what they watch and when they watch it on their main set free from subscription.

Panasonic will also introduce Freeview Play products including three Blu-ray recorders and two digital TV recorders, with recorders in the pipeline from Humax, Manhattan TV and Vestel, and and other major TV manufacturers to follow.

Humax Freeview Play box on sale early

In October Humax placed the first Freeview Play set-top box in shops a week earlier than expected, giving catch-up access to BBC iPlayer, ITV Player, All 4 and Demand 5 alongside three Freeview HD tuners.

The FVP-4000T comes with 500GB or 1TB recording capacity, and has built-in WiFi. See page 20 for further information.

YOUVIEW

PlusNet launches YouView service

Sheffield-based internet provider PlusNet has quietly launched a YouView service, which is designed to be taken as part of a bundle alongside its ISP's residential Unlimited Fibre Broadband and phone package. Assuming you already have one of the ISP's fibre broadband and phone line packages then for an extra £5 per month you can take out their TV package – this includes the Entertainment channel pack with 15 premium Entertainment channels.

Sony collaborates with YouView on new TV range

Sony became the first TV manufacturer to offer in-built YouView through its 2015 Bravia range, giving users the TV platform without the need for a separate set-top box.

The inclusion of YouView on the new Bravia range will be combined with Android TV and Sony's One-Flick Entertainment, offering an integrated Smart TV experience with access to a wide range of content – including demand services, movies, music, photos, games and apps.

Customers will be given access to YouView's extensive library of more than 12,000 on-demand films and TV programmes.

SKY

New 'Sky Q Silver' set-top box

In November Sky announced a new range of products including the 'Sky Q Silver' set-top box. Sky Q products will offer a blend of live and on-demand TV including 4K Ultra HD content. The boxes will have 12 channel tuners. These will let you record up to four programmes at a time while watching a fifth. They will also be able to stream content via WiFi to smaller Sky Q Mini boxes feeding up to two secondary TV sets around the home.

The service will launch in early 2016, but prices have not yet been announced. This is the most significant new product launch for Sky since it rolled out its HD packages a decade ago. Sky will also launch an Ultra High Definition service in 2016. The new service will offer customers a range of sports, movies, and entertainment content with up to four times the detail of HD. This may not be ready by the time the Sky Q Silver set-top boxes launch. See page 28 for further information.

Sky refines Sky+ EPG

Earlier this year Sky refined its programme guide giving more prominence to customers' recent recordings and downloads, adding them to the top of the first page, from their previous placing at the end of the planner.

Planned recordings are now in a new 'Scheduled' page alongside Smart Series Links which not only record the current season, but also remembers to record all episodes of future seasons when they come back on TV. Sky has previously invited customers to register their favourite shows online, to be alerted when a new series makes its debut.

An updated 'Family' setting hides all adult channels, pin protects all pre-watershed playback of on-demand content and movies with a 12 rating and above.

The Eco Mode has been extended by 50%, now automatically switching off the Sky+HD box from 2.45-5.45am to save more energy.

Sky to add subtitles to online on-demand content and box sets

Sky is responding to changes in the way audiences view programmes by adding subtitles to its internet delivered on-demand services and box sets by summer 2016.

Currently subtitles are available on Sky's regular, linear broadcasts transmitted via satellite, which can be recorded and played back with subtitles. Subtitles are also available on those on-demand programmes that are delivered via the satellite signal and found in the Top Picks section of the Sky TV guide.

Sky+ personalises movies experience

Sky has added more personalisation and discovery features to the Movies section of Sky+, with a personalised Watchlist where viewers can compile a wish list of films.

The More Like This provides recommendations through the yellow button as you browse the Sky Guide, and Watch From Start instantly launches a film via on-demand when you see it on a live Sky Movies channel.

Now TV - Sky offers no-contract monthly subscription to Sky Sports

In August Sky began offering monthly subscriptions for its sports channels without the need to subscribe to any other Sky TV package. Users of Sky's Now TV internet streaming service can take up the new Sky Sports Month Pass for £31.99 a month.

There are no contract periods for Now TV's services and customers will be able to

NEWS

take out a single month's subscription. Day and weekly Sky Sports Passes will continue to be available for those who just want Sky Sports for a short period of time. The Now TV offer includes all seven Sky Sports channels: Sky Sports 1, 2, 3, 4, 5, F1 and Sky Sports News HQ.

Now TV: New box, new interface
Now TV launched a new box in August. Retailing at £14.99, the new box will include an Ethernet port, so won't require a WiFi connection to the internet. The new box also includes a processor that is five times faster than that with the previous model.

FREESAT

Humax delivers Freesat compact and connected
In July Humax launched a new compact set-top box for free-to-air satellite platform Freesat.

The new HDR-1100 range all have built in WiFi – earlier models needed to be connected over Ethernet – and included access to catch-up services branded as Freetime. These include BBC iPlayer, ITV Player, YouTube and Curzon Home Cinema.

The Humax Freesat HDR-1100S comes in a glossy black or white finish. It is priced at £189 (500GB), £219 (1TB) and £299 (2TB).

Freesat reveals LG partnership
LG will launch a new range of Freesat-enabled smart TVs and upgrade some existing models to work with the free-to-air satellite services.

The new LF650 and LF630 TVs will be sold in screen sizes from 32in to 55in, while a software rollout will put the Freesat EPG on existing TVs with satellite tuners built in.

VIRGIN

Virgin Media sticks with Tivo until at least 2018
Virgin Media has extended its PVR and cloud TV services partnership with TiVo until 2018. The three-year extension to the original 2010 agreement commits Virgin to 'the development of next-generation solutions from TiVo'.

TiVo boxes are now installed in more than 2.5 million Virgin Media customer homes – over half its TV customer base and TiVo technology is also behind the Virgin TV Anywhere service launched in 2012.

New Tivo Box can skip adverts, record and stream 4K video
TiVo has released a feature-packed new digital recorder box which can skip adverts, record four different shows at once and stream from Netflix and Amazon Prime Instant Video in Ultra HD 4K. It's called the TiVo Bolt.

Currently only available in the USA the most interesting feature of the TiVo Bolt is its ability to skip through an entire ad break at the touch of a button. The box tags the start and end of commercial breaks so viewers can skip that section when watching on their recordings.

At launch, the feature works with some 20 over-the-air and cable channels, including the major broadcast networks, mostly during prime-time hours.

The Californian company hopes to avoid any legal problems with its commercial hopping features – dubbed SkipMode – by leaving it to users to enable the feature.

In addition to this, and for those really short on time, the TiVo Bolt also has a QuickMode that plays back recordings 30 percent faster. The pitch on the audio is adjusted using software so that it won't sound odd. With this and the commercial skipping, it will be possible to watch an hour-long show in roughly a half-hour.

Other features include support for video in Ultra HD, or 4K, resolution when channels start offering that quality.

Virgin Media already uses TiVo hardware for their UK TV customers – and it will be interesting to see if the company offers the latest TiVo Bolt hardware to its customers.

BT

BT TV announces pricing for sports coverage and 4K
BT began offering football and rugby matches in 4K quality in August 2015. BT Sport Ultra HD became the first 4K channel in Europe. Ultra-high definition, or 4K, offers four times the resolution of current HD.

Television Viewer's Guide

NEWS

The matches will be free for customers who subscribe to BT TV, while BT broadband customers who choose to watch via Sky or the BT Sport app will be charged £5 a month.

To use the service customers will need a BT 4K-compatible set-top box, and a 4K-compatible TV. This service will only be available to BT Infinity customers, due to the bandwidth required to stream 4K.

BT buys mobile provider EE
BT has spent £12.5bn acquiring mobile provider EE, which was founded in 2010 by the merging of Orange and T-Mobile. The deal will enable BT to offer 'quad-play' contracts. Quad-play means bundling TV, broadband, mobile phone and home phone services into one big contract.

The group plans to sell a full range of its services to the combined EE/BT customer base, including BT offering broadband, fixed line and pay-TV services to EE customers who do not currently buy a service from it. It is also expected to offer mobile deals to its own customers who take one or more of its services.

It is likely to be particularly keen to target customers with a bundle deal – all these services in one package. Talk Talk and Virgin Media already do this. EE has recently launched a TV service and it is not yet clear how the deal will impact on EE TV.

EE

EE TV replay option annoys commercial broadcasters
Some of the UK's biggest commercial broadcasters have come out against a function offered by EE TV, which allows viewers to skip advertising.

EE TV offers viewers a replay option, which enables viewers to catch-up on programmes recorded onto the hard drive of their EE TV box.

Crucially, the programmes, which are automatically recorded in the background, are then available via a special menu which works in a similar manner to video-on-demand services.

Viewers are free to skip adverts during the recordings, in contrast to many on-demand services, which feature adverts that have to be watched at the beginning and during programmes.

ITV Player and All 4 are not available on EE TV, and the replay function offered to viewers is a major hurdle preventing new catch-up services being added to EE TV. Even Demand 5, which is available on EE TV, could risk being pulled from the service over the replay issue, with Channel 5 reportedly having major reservations.

TALKTALK

TalkTalk releases TV2Go app
In February TalkTalk launched TV2Go, a new app that allows customers to watch live channels and on-demand content on the go when connected to WiFi, amid growing investment in its TV offering.

TalkTalk's new app aims to provide a seamless viewing experience for subscribers, as the app lets you pause on-demand shows on the YouView box and pick up from where you left off on the app and vice versa.

TalkTalk has said that it is the first quad play – broadband, TV, mobile and phone – provider to offer this functionality. TV2Go also includes a feature that plays all episodes in series one after another.

BBC IPLAYER

BBC Global iPlayer closes
The BBC 'global' version of its iPlayer on-demand service closed in June this year.

The corporation had charged users subscription fees to watch programmes via the app in Western Europe, Australia and Canada, although it never reached the US.

Reports suggest American pay-TV operators in the USA had threatened to drop the BBC America channel if the app had launched locally because they believed it would cost them viewers.

The content provided was distinct from that offered via the UK version, including older shows from the corporation's library in addition to recently broadcast programmes.

BBC iPlayer moves away from Flash and towards HTML5
The BBC's iPlayer has been made available using the HTML5 web language, at the

NEWS

expense of Adobe's Flash player.

The broadcaster's media service was one of the most prominent online platforms to use Flash.

Adobe's plug-in has been criticised by some security experts, who said it was a weak point of many sites. The BBC has said that it was now confident it could achieve high playback quality without using a third-party plug-in such as Flash player.

Users have been invited to visit a BBC site where they can set a cookie in their browsers that will allow them to access the HTML5 player when they visit iPlayer in future. However, the Flash version will remain available.

Restrictions to overseas access to BBC iPlayer

The BBC has begun to crack down on commercial virtual private networks (VPNs) that are sometimes used to stream the iPlayer to overseas destinations, in breach of its terms of use.

Almost 65 million people outside the UK are bypassing restrictions to view the BBC iPlayer using proxy servers and virtual private networks (VPNs).

iPlayer: New features added to BBC on-demand service

The BBC has been rolling out new features on the iPlayer, as part of its drive to 'personalise' the user experience. The on-demand TV service is changing to keep pace with popular streaming services such as Netflix.

The service is rolling out a Live Restart feature – which already works on desktop computer versions of the app – to smart TVs, enabling users to jump back to the beginning of a show at any time during the live broadcast.

Cross-device pause and resume – the ability to pause a stream on one device and then pick it up on another – has also been added, as has the My Programmes section, which already works on the iPlayer website, but will now expand to mobile devices. The feature stores favourited and watched shows in a single, personal section of the app.

Audio Description (AD) functionality is now available on iPlayer on TV platforms – as well as computers, mobiles and tablets.

This latest update comes after a major redesign last year, which was aimed at making iPlayer feel more like television.

NETFLIX

Netflix coming to more TVs and set-top boxes

Smart TVs and other networked home entertainment devices from Panasonic, Philips, Sony, Toshiba and Vestel in Europe are starting to have a Netflix button on their remote controls. This enables you to access the service without navigating through the smart TV menu.

Humax has also signed a deal to add Netflix to many of its existing devices as well as future products – although it has yet to confirm which devices and which territories are covered.

AMAZON

Amazon Fire TV now supports 4K

Amazon's new Fire TV box is the company's 4K UHD-capable follow up to its successful streaming media box. It costs £80 and was released in early November.

It has few competitors capable of streaming 4K video available in the UK. The Roku 4 is capable of 4K video streaming costing at least £100, but had not been released in the UK at the time of writing. The new box may be worth considering if you subscribe to Amazon's Prime service.

Of course, in order to watch 4K content, users will need a 4K TV, but check first as it is not compatible with all 4K TVs.

In addition to the 8GB of built-in storage, the new Amazon Fire TV box now offers expandable storage using a microSD card slot, for up to 128GB of additional storage for downloaded apps and games.

Fire TV is available from Amazon, Dixons, John Lewis, Argos, Tesco for £79.99. An additional game controller with voice search, refined controls, a headphone jack and up to 90 hours of battery life is available for an additional £39.99.

Amazon's new remote control for Fire TV Stick

In September Amazon announced a new

remote control for its Fire TV Stick – the compact version of its Fire TV that looks like a USB stick. Earlier in 2015, Amazon announced the Fire TV Stick was its fastest-selling UK device ever.

The voice remote allows users to speak the name of a film, TV show, actor, director, or genre and see the results pop up on the screen, bypassing the tedious task of spelling out words using an on-screen letter grid.

Fire TV Stick with Voice Remote costs £44.99. Customers can purchase the Fire TV Stick on its own for £34.99.

GOOGLE

Google Chromecast Audio device

The Chromecast Audio is Google's wireless streaming dongle that turns almost any speaker into an internet-connected one. The small black puck is very similar to Google's standard Chromecast device, but the audio version has lost the HDMI cable and gained direct analogue or digital audio output.

You set it up using the Chromecast app on an iPhone or Android device. Plug the supplied USB power adapter and cable into the Chromecast Audio, plug in the included 3.5mm audio cable (alternatively, a 3.5mm optical cable [Toslink] or 3.5mm cable to phono audio lead), hook it up to your speakers and fire up the app on your smartphone or tablet.

The one requirement that might catch some out is that the Chromecast Audio needs a WiFi network, so it's not ideal for use out of doors with battery powered speaker systems. Neither is there an Ethernet port.

Streaming music to the Chromecast Audio is easy and should be a familiar experience for anyone used to using apps on their smartphone. The WiFi device supports any audio streamed over the Google Cast protocol. That includes apps such as Spotify, Google's Play Music service, media locally stored on the smartphone and system audio mirrored from your Android smartphone or tablet. It doesn't work with iTunes/Apple Music.

The smartphone then becomes a remote able to skip tracks, select different albums and change the volume. If streaming music from the internet rather than locally, Chromecast Audio connects directly to the service and doesn't rely on the smartphone and its battery to continue playing music.

Given that wireless speaker systems are becoming increasingly popular, but can also cost hundreds of pounds, the £30 Chromecast Audio looks like an affordable way to dabble in music streaming without having to replace your old sound system – or even to save an old unused speaker from simply gathering dust.

Google Nexus Player

The new Google Nexus Player, £80, works in a very similar way to the Chromecast dongle, but it offers slightly higher specifications and improved WiFi performance. Its support for 1080p means you should see significantly better video quality.

Like Amazon's Fire TV, the Google box can be paired with a game controller to take advantage of the wide range of games for Android. The optional controller is £34.99. Unlike the Amazon Fire TV which has a dedicated BBC iPlayer app, on the Google Nexus Player you have to send this content from another device, such as smartphone, tablet or laptop – it's easy to do, but not quite as simple as having a built-in app, as on the Fire TV box.

The included remote control has voice control too, so you can search for things as you would do on a smartphone using voice commands.

Google's new Chromecast device

The original Google Chromecast device has been a great success. Costing £30, the compact device gave owners a low-cost way of upgrading an old TV into a smart one, so they could stream content from the likes of BBC iPlayer and Netflix using a phone or tablet as a remote control.

The new Chromecast, £35, comes with a change of style. Instead of a solid stick Google has opted for a disc-like design available in black, red or yellow. It connects to your TV using a short, flexible HDMI cable, rather than plugging in directly to an HDMI socket like the old device.

The new device offers better performance. It has a faster processor, and WiFi connectivity has been improved. There's

NEWS 13

also a new 'fast play' feature, which Google claims speeds up connecting your mobile device to the Chromecast, so content is displayed on your big screen quicker when you hit the cast button.

Google has also given the Chromecast app an overhaul, making it more user-friendly in the process. The new app brings all the content that you may want to watch into one place, so you don't have to trawl through all your individual apps to find the programme you're looking for. It will only display content that's available via apps you already have installed and that are compatible with the cast feature.

APPLE

Apple subscription TV service facing further delay

Viewers hoping to get their movies and television shows over an Apple-administered TV service will have to wait until at least 2016.

The much-hinted-at subscription TV service was expected to be unveiled in 2015, but plans for the September launch have been scrapped because of a lack of progress in securing licensing deals for programming from content providers such as TV networks.

Delivering TV programming to viewers is widely expected to be the next step in Apple's entertainment strategy, a key component to Apple's overall goal of creating a stable of products and services that keep consumers from straying to competitors such as Google. The company revamped its music strategy in May with the addition of streaming-music service Apple Music and has been working hard to attract more users and retailers to its Apple Pay, the mobile-wallet platform Apple launched in October.

The delay highlights the challenges that even a major player like Apple faces when introducing a video service that seeks to upend a crowded industry already dominated by cable companies, satellite TV providers and even streaming services such as Netflix and Amazon Prime. Because of that competition, content providers can hold out for better terms.

New Apple TV box

In September Apple CEO Tim Cook unveiled the new Apple TV device, ten millimeters taller, with an updated remote control, and a built-in App Store. The box however does not support 4K video.

'It's the golden age of television, but the television experience has been virtually standing still', said Cook. 'Today, we are going to do something about that. It starts with a vision. Our vision for TV is simple and perhaps a little provocative. We believe the future of TV is apps.'

The new remote control and its tvOS operating system supports Siri. This is Apple's 'intelligent assistant' that enables users of Apple devices to speak natural language voice commands in order to operate the mobile device and its apps – meaning that the box should be fully operable just with your voice.

Launched in late October the 32GB version costs around £96 while the 64GB version will come in at about £129.

TVPlayer launches premium service

TVPlayer has launched a premium add-on service, TV Player Plus, that enables viewers to access extra channels. These include Discovery, British Eurosport and Gold, and are streamed to their TV, computer or mobile device using the app.

TVPlayer Plus will be priced from £4.99 per month (with no contract) and enables live streaming of a range of premium television channels that are not available on BBC iPlayer. It is available on iOS, Android, Amazon Fire TV, Mac, PC, selected Smart TVs, the Amazon Fire TV Stick and Google Chromecast. Existing users of TVPlayer can register for a one-month trial of TVPlayer Plus.

TVs soundbars and equipment

Last year saw retailer John Lewis enter the TV market with its own-brand 1080p TV, the 55JL9000. In June this year the company introduced 4K models.

Three models are currently available in its JL9100 range. The 40 inch 40JL9100, £529; the 49 inch 49JL9100, £679; and the 55 inch 55JL9100, £899.

Television Viewer's Guide

Ultra HD/4K and HDR

New and enhanced TV systems are coming onto the scene. If you're looking for a new television then the chances are that you've heard of 4K, Ultra HD TV. This and HDR (High Dynamic Range), are two of the main features that manufacturers are promoting this year. We look at what they offer and whether you should be investing just yet.

We first looked at 4K televisions in the guide two years ago. Also known as ultra high-definition (UHD) or Ultra HD, 4K is the next step-up in resolution for flat-screen displays, offering around four times the pixel resolution of a 'regular' 1080p HDTV. Since then technology has advanced and most manufacturers have 4K TVs in their range.

What does the technology offer?
Officially, 4K resolution is 4096 x 2160 pixels. However, in order to shoehorn this higher resolution video on to a normal 16:9 picture format, it has been altered to 3840 x 2160 – still four times the total number of pixels on a Full HD 1080p screen (1920 x 1080). The most obvious benefit of 4K TVs is that their pictures have the potential for more detail and sharpness. However whether you will see the benefit depends to a large extent on the size of the screen and how close you will sit to it. Put simply, if you are viewing a small 4K screen from a distance you are unlikely to see the benefit.

In order to take full advantage of 4K Ultra HD you will, of course, need a compatible TV, a source and content that packs those all-important extra pixels.

HDMI connections
Early UHD/4K televisions were restricted in the quality of their output by older HDMI 1.4 connections. The arrival of HDMI 2.0 has enabled 4K resolution at higher frame rates, namely 50p and 60p. This is particularly useful for fast-paced, action-packed programmes such as live sports broadcasts.

What 4K content is available?
One of the big issues for early adopters of 4K has been the lack of content. This is now beginning to be addressed, but primarily by a limited quantity of internet-delivered content from the likes of Netflix, Amazon and BT TV. BT's new Ultra HD/4K channel launched in August, and showed the FA Community Shield match between Chelsea and Arsenal. It's a subscription channel and viewers will need to purchase a BT Infinity fibre-optic broadband package to use it.

Broadcast 4K content
Broadcasters are testing live 4K content. It's not something that we are likely to see on Freeview, due to the large bandwidth requirements, but satellite and cable services will certainly embrace it.

So far, Sky UK has been tight-lipped over its Ultra HD launch plans. The company is rumoured to be about to launch a new set-top box, dubbed 'SkyQ'. Like the new BT YouView box it will be capable of 4K reception.

4K Blu-ray
The specification for 4K Blu-ray has been agreed. The new Ultra HD Blu-ray standard will enable the high capacity discs to deliver content at the 4K resolution of 3840 x 2160 with an upgraded colour range. New 4K Blu-ray discs will enable high dynamic range (HDR) content as well as offering high-frame rate (HFR) playback. Ultra HD Blu-rays should be on sale in the UK by Christmas 2015.

Technologies that work with 4K
Although 4K will offer better pictures for viewers with large screens and a source of 4K content, it is the associated technologies that will be included with 4K TVs that may prove to be of more importance to most viewers – in particular those with 'normal-sized' televisions. The two most interesting developments are Quantum Dot, and HDR (High Dynamic Range).

ULTRA HD/4K AND HDR

Quantum dot
This is an additional technology that can be utilized by 4K TV sets. Both Samsung and LG have begun to introduce quantum dot technology to their screens. The system uses crystals more than 50,000 times smaller than the width of a human hair that can generate vivid colours and are capable of being more than twice as bright as conventional TV screens.

What is HDR?
Like Quantum dot, HDR, or High Dynamic Range, isn't a competitor to Ultra HD/4K, but rather an additional technology that will become available on some 4K TVs and will improve the pictures you see on screen.

The term HDR is already used in the world of photography, where HDR images are an option in high-end stills cameras and smartphones.

An HDR photo usually combines several images taken at different exposures into a single picture that appears to have even more detail and sharpness. The system works similarly with video in compatible HDR TVs, creating a greater dynamic range between the blackest blacks and the whitest whites, plus all the subtle tones in between.

How do HDR images look?
This means that TV pictures will seem even more detailed, particularly on darker scenes. Images also have improved colour fidelity, nearly approaching those found in nature and better matching the capabilities of the human eye.

HDR pictures are punchier, richer and more 'luminous' than ordinary ones, and unlike UHD images do not demand a very large screen to be effective. HDR brightness for home screens is likely to be about double that available in today's models; more than that (depending on the ambient light level) could be uncomfortable and distracting. Particularly striking in an HDR picture is the enhanced colour rendering in dark areas.

Standards and specifications
HDR systems are being developed by Dolby Vision, Technicolor and Philips. Both colour and luminance (brightness/contrast) have higher bit-rates than at present, along with other tweaks; the BBC has proposed that HDR standards be added to the current DVB-T standard.

There have been concerns expressed regarding the bandwidth demands for HDR content delivered across the internet. Netflix, one of the key suppliers of 4K content in the USA, is suggesting that there will be an additional 20 percent bandwidth requirement for HDR on top of Ultra HD content.

If you are looking for a receiver box for HDR make sure it has a suitable HDMI port: v2.0a is the minimum requirement here.

TV models supporting HDR
Some current TVs are already HDR-compliant. For instance the Samsung 8 and 9 range, amongst which type UE65JS9000 gives stunning pictures.
Sony models X93C and X94C, also the new 65S8505C have HDR support, and Sony will enable it in its existing X85C/90C/91C/93C/94C models via a network update.
LG's models LF55/65EF9500 support HDR, while their previous curved OLED screens are updateable for web streaming but not for HDMI operation.

Certain late Panasonic TVs will be made HDR-compatible with a firmware update.

Receiving broadcast HDR TV
No over-the-air satellite or cable broadcaster offers HDR at present; however once the standards have been set and agreed it's likely to go hand-in-hand with UHD/4K services.

HDR via the internet
For the present HDR pictures come via the internet, and as already mentioned require a respectable data rate. Amazon provides HDR content free to Prime subscribers. Netflix is expected to start an HDR service in 2016, though details were scarce as we went to press.

If you are looking for a new television set, and aren't in the market for a huge screen we feel that the benefits of 4K will be marginal. Instead our recommendation if you are looking for the best picture quality at a reasonable price is to look for sets featuring the HDR feature.

WHICH DIGITAL PLATFORM TO CHOOSE

Most viewers now have several digital TV options. Hybrid systems, such as YouView, where the main channels are carried by Freeview with additional and catch-up content delivered by broadband, are increasingly common.

Freeview / Play

Freeview is the UK's terrestrial TV system.

Freeview carries around 60 TV, 24 radio and up to 12 HD channels (depending on where you are in the UK).

Freeview Play launched in 2015. It offers Freeview and catch-up TV from BBC iPlayer, ITV Player, All 4 and Demand 5, and it is subscription free.

Equipment
All new televisions have a Freeview tuner, but to receive all channels look for products with an HD tuner.
For Freeview Play you will need a new Freeview Play TV or Freeview Play recorder to use the service.

Pros and Cons
If you get a good signal Freeview is a good TV option - and there are no subscription charges. The range of channels is not as wide as that available on Sky, for example. You can add to what's available by opting for YouView, by using a Smart TV, or by accessing paid for content via an increasing range of online options.

Note that if your aerial points at a relay, rather than a main, transmitter, you will not receive the full range of Freeview channels.

Sky and NOW TV

Sky provides a good range of channels including sports and movies, and subsidised installations. The Sky+ box set a high standard in TV recording, but is now being matched by similar systems from Freeview and Freesat. Limited 3D content is available via Sky's On Demand service.

Sky sells a range of television subscription packages, but you do not have to pay a monthly subscription if you simply want Sky's free channels. The FreesatfromSky scheme offers a highly competitive package with a dish, digibox and installation - all for £175 and with no monthly subscription.

Equipment
You'll need a satellite dish and a set-top box. **NOW TV** is Sky's internet service and with a NOW TV box and internet connection, you can pay for Sky's content on a daily, weekly or monthly basis.

Pros and Cons
Good UK signal coverage, and a great range of channels including sport and movies. On the downside, you'll have to pay a subscription for all but the basic channels.

Freesat / Freetime

Freesat offers a good alternative to Sky, as well as filling in the gaps in Freeview's terrestrial coverage. It carries over 200 TV and radio channels. Unlike Sky it offers subscription-free HD television, with a range of HD channels carried. Catch-up content is available from BBC iPlayer, ITV Player, All 4, and Demand 5 on Freesat's 'Freetime' equipment and some HD set-top boxes.

Equipment
You'll need a satellite dish (you can use a Sky dish) and a Freesat set-top box. Panasonic is one of the few manufacturers that sells televisions with built-in Freesat tuners. HD boxes start from £50. PVRs that record (Freesat+) start at around £140. Freesat Freetime boxes offer similar functionality to YouView.

Pros and Cons
Good coverage and subscription-free recording, plus free HD and catch-up TV with Freetime. Installation is simple - particularly if you already have a Sky dish. The range of equipment is improving, but Freesat could do with more HD channels.

WHICH DIGITAL PLATFORM TO CHOOSE

Virgin (Cable)

If you are fortunate enough to have the service running past your door, cable TV is worth considering. Virgin is the main provider. There is a good range of TV and radio channels available, with cable offering other benefits, such as telephone, and some of the fastest broadband packages.

On the television side, cable offers video-on-demand at much faster speeds than available on broadband, plus a good range of channels. Catch-up TV from BBC iPlayer, ITV Player, All 4 and Demand 5 is available on Virgin set-top boxes.

Equipment
TiVo boxes are offered as standard, for which you'll pay £5 per month. The basic TV M package (60+ channels) is only available as part of Virgin's Big Easy Bundle. This costs £22 per month plus a £50 installation fee and £17 monthly Virgin phone line rental. The basic M+ TV package costs £21 per month.

Pros and Cons
A good range of channels, and catch-up TV, plus additional phone and very fast broadband services are Virgin's main attractions. The fact that there is limited UK coverage, the need for a monthly subscription and 18-month contracts will put off some viewers.

BT TV

BT TV is a Freeview-based system with additional catch-up and on-demand content delivered by a YouView set-top box. It's only available to BT Broadband customers.

The basic TV Essential package plus broadband costs £5 per month plus phone line rental of £18. A TV Entertainment package offering 20 additional channels, along with a faster BT Infinity fibre-optic broadband connection, costs £35 per month, plus line rental of £18.

BT Sport is available free with BT Infinity packages. If you want Sky Sport you'll need the older BT Vision equipment, and it will cost from £22 per month.

Equipment
BT provides a 'free' Humax DTR-T2110 to new customers. You'll also be provided with a BT Home Hub wireless router. You may need Powerline plug adapters if your router is beyond the WiFi range of your TV.

Pros and Cons
BT TV offers useful PVR recording functions and access to catch-up from the likes of BBC iPlayer, and on-demand content and paid-for film downloads.

On the downside, you are tied into a 12-18 month contract with monthly fees.

YouView

YouView offers a hybrid digital TV service with content from Freeview plus additional content delivered via broadband.

You can get YouView directly from a retailer for a one-off payment for the set-top box, or from broadband providers BT TV, TalkTalk or Plusnet. If you are not interested in paid-for, or premium content, we'd suggest that you choose Freeview Play instead.

Catch-up content is available from BBC iPlayer, and other broadcasters; with paid-for content from NOW and YouView providers.

Equipment
The BT badged Humax DTR-T2110 is available for £150. The Humax DTR-T2000 1TB box can be purchased outright from John Lewis for £230. Subsidised boxes are available if you subscribe to a YouView package. You will need Freeview and reasonably fast broadband connection.

Pros and Cons
Subscription-free HD and on-demand services plus recording/PVR features. Watching on-demand may count towards your monthly broadband data allowance, so it may pay to choose a service from a YouView provider.

Television Viewer's Guide

Freeview

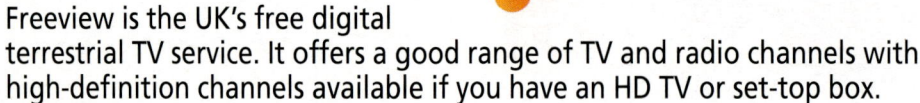

Freeview is the UK's free digital terrestrial TV service. It offers a good range of TV and radio channels with high-definition channels available if you have an HD TV or set-top box.

The main news for Freeview this year was October's launch of Freeview Play, a service that combines Freeview channels with catch-up internet content. In addition the Freeview logos have been redesigned, and Freeview+ has been renamed FreeviewHD Recorder. From 2016 only HD products will carry the Freeview trade mark.

What does Freeview cost?
There is no contract to sign and there are no monthly subscription charges to use the standard Freeview service. You will only have to pay subscription fees if you wish to watch additional pay content from providers such as TalkTalk, BT TV and Plusnet who provide paid for content alongside Freeview channels in a YouView package.

What equipment do you need?
Freeview use requires an aerial. In good signal areas you may get away with a set-top or loft aerial.

All new televisions have a built-in Freeview tuner. If you have an older analogue television, you will need a Freeview set-top box (also called a receiver, digibox or STB). Set-top boxes start from around £20, with digital TV recorders (PVRs) available for around £100.

You will also find Freeview tuners built into devices such as Blu-ray players and recorders. Freeview is also used to deliver the main channels in **YouView**, which combines Freeview channels with premium and online catch-up television content. This is covered in more detail in the YouView section in the guide.

Do you need a new TV?
All new television sets work with Freeview. If you have an older analogue television, a cheap Freeview set-top box will enable it to receive digital signals. A new set-top box can also be useful if you want to add high-definition or Smart functions to an older television.

What channels are available?
Freeview carries over 60 TV channels and 25 radio stations. Around 12 channels, depending on your location, are available in HD – all are subscription free.

Are you covered?
The last remaining analogue TV transmitter was switched off in October 2012 – so with the exception of some remote areas and pockets where there is no signal you should have coverage. You can check if you enter your postcode on Freeview's website: www.freeview.co.uk

If you cannot receive a Freeview signal due to interference, or screening by hills or tall buildings, we'd recommend you look at Freesat as an alternative.

Will you receive all the Freeview channels?
Freeview Lite is a term used to describe the limited service that is available to viewers who pick their signal up from a relay transmitter, as opposed to a main transmitter. Most relay transmitters only carry three of the six available multiplexes. You will still receive the main channels, but will not be able to receive channels such as Dave and Film4.

What are multiplexes?
The channels carried on Freeview are bundled together in what are called multiplexes. The Freeview digital tuner separates the multiplex into individual channels which are then listed in the programme guide. Multiplex details for channels are listed on page 44.

How do you pick up HD channels?
HD channels include BBC One HD, BBC

FREEVIEW 19

Two HD, ITV HD, Channel 4 HD and S4C HD in Wales. To watch HD channels you will need a TV or set-top box with an HD-compatible Freeview tuner. HD boxes often include additional features such as access to BBC iPlayer and other Smart TV features.

Interference from 4G
A small number of viewers have experienced 4G interference to their TV signal as 4G mobile networks have been rolled out.

FreeviewHD Recorder
Freeview recording products are now badged with the logo **FreeviewHD Recorder**. Prior to this they were called Freeview+. The name may have changed but the functionality remains the same.

Products offer the ability to set up recordings easily, pause live TV for up to 30 minutes, and offer an electronic programme guide (EPG). Freeview recorders (PVRs) have their own EPG layout offering information about the coming 7-14 days programmes - some are better than others. Most are simple to navigate, clear to read and display several channels on screen at once.

PVRs offer the facility to watch one channel while recording another, can record a whole series easily, and will warn you of clashes in recordings.

Disk size
Each gigabyte (GB) of storage holds over half an hour of standard-definition television, so for instance, a 500GB disk is good for 250 hours of programmes. As a rule of thumb divide the capacity in GB by two to get the SD storage capacity in hours. Most PVRs can also record radio programmes.

Features to look for
The key thing to consider when buying Freeview equipment is whether you need SD (standard-definition) or HD (high-definition) equipment. HD gives more flexibility and usually offers access to services such as BBC iPlayer and ITV Player – provided you have a broadband internet connection.

If you are choosing a Freeview television set or set-top box ensure that it has the necessary Scart, HDMI and AV connections.

Recommended equipment
All new TVs have a built-in Freeview tuner, with most medium and larger sized screens now offering HD. By 2016 all Freeview-badged products will need to offer HD,

If you are looking for a set-top box to convert an old analogue TV, you can expect to pay around £20 for a basic model and up to around £60 for something with HD and more advanced features.

Freeview boxes that record start at around £100 for a basic SD model rising to up to £200 for an HD recorder. We've listed a few of the models that we like below.

Humax FVP-4000T £200
Launched in October 2015, this is the latest Freeview Play set-top box. It offers three HD tuners, WiFi, Ethernet and DLNA connection and dual USB ports. You'll find a review of this new box on page 21.

Humax DTR-2000 £175
This YouView compatible box seems to be replacing the older DTR-T1010 model. It offers the usual twin tuners and a 500GB disc. For an additional £55 you could upgrade to the 1TB model. There's no WiFi built-in, but you can use an Ethernet cable, or the HomePlug internet-over-mains system to connect to a remote router.

Hard disk and DVD combinations
These offer both functions in one box. One advantage of this is less space taken under your TV; another is fewer cables to connect. They also offer fast copying from hard disk to DVD – useful if you want to share or archive your recordings. Both Panasonic and LG offer a range of combination machines, some of which include Blu-ray.

Freeview Play
See the following pages for information about Freeview Play and the recently launched Humax FVP-4000T set-top box that works with the new subscription-free service.

Television Viewer's Guide

Freeview Play launched in October 2015. It is Freeview's next generation smart TV platform, and was originally called Freeview Connect. It offers live viewing and catch-up TV on the traditional terrestrially delivered service, Freeview. Importantly it is subscription-free and works on all existing broadband services.

Freeview Play is the new take on Smart TV from Freeview. It offers a mix of live and catch-up TV with the convenience of an EPG that scrolls back seven days as well as forwards. This is the critical feature that separates Freeview Play from many other Smart TV PVRs that have Freeview HD tuners and separate catch-up TV portals, but it's also very similar to YouView.

Freeview Play boasts around 60 Freeview channels, including 12 HD ones – essentially the same channels you find on the standard Freeview service. As well as these, you'll also have access to catch-up services such as BBC iPlayer and ITV Player.

The goal is to make Freeview Play an open platform available on a wide range of products, but it is currently being offered on just a few devices that support Freeview HD and internet 'smart' features. Freeview Play ready products will display the new logo.

Freeview Play – can I get it?
Freeview is used for the mainstream TV channels, so if you're in an area with good Freeview reception you will be fine. You will also need to connect your TV to your broadband router. All the Freeview Play products launched at the time of writing, have built-in WiFi making connection simple.

Freeview Play - key features
Freeview Play offers a fairly standard EPG. Enabling you to scroll forward and backward, when connected to the internet, up to seven days to see what's coming up, or what you've missed. You can navigate the schedule via date and get information on what each programme is about. If your device has recording functionality, you can also select programmes to record via the guide.

Catch-up TV
Connect your Freeview Play TV or set-top box up to the internet and you can scroll through the electronic programme guide backwards to catch up on programmes you've missed over the past week. BBC iPlayer, ITV Player, All 4 and Demand 5 are all on offer, and content from each is displayed by day for easier browsing.

Channels such as Dave, Yesterday and Really are excluded from the Freeview Play EPG and don't expect to find everything that's shown on the likes of BBC One, BBC Two and ITV to be available as many movies, dramas, sports shows and news programmes are excluded because of rights issues. That said, an awful lot of stuff is available from the 30-odd channels that fall within the umbrella of the big four UK terrestrial broadcasters – BBC, ITV, Channel 4 and Five.

You can also access all of the catch-up content via separate on-demand portals or apps for each provider, where you'll also find YouTube, BBC News, BBC Sport, internet radio and a slew of other minor radio stations, news, lifestyle, religious and cooking channels. There's no Netflix (at launch) but this is promised for 'early' 2016. Apps are accessed from a separate section within the . Plus you can get the likes of Netflix, Amazon Prime and YouTube from most TVs and PVRs.

Television Viewer's Guide

FREEVIEW PLAY   21

Humax FVP-4000T, £200 for the 500GB and £230 for the 1TB version.

Humax FVP-4000T

Equipment-wise, the Humax FVP-4000T is the first PVR with Freeview Play built-in. The box itself is attractive, especially in its beige and white livery with brushed metal trim (a brown version is also available). It's compact for a PVR (280 x 48 x 200mm) and comes with a decent-sized remote control that seems logically laid out. Two configurations are available – offering 500GB (£200) or 1TB (£230) recording capacity.

Set up is helped by a step-by-step wizard which includes the option to display menu text in English, Gaelic, Irish or Welsh.

The EPG is big and brash, with lots of information for each show, plus (most of the time) a screenshot to provide an instant visual identifier. There's no live window showing the current live broadcast but you can create a list of favourite channels to minimise the amount of time required navigating the EPG.

Recording and viewing saved shows is intuitive. Recordings are stored in a tab, where they're sorted by title, channel and the day of the week they were recorded on. The 500GB model provides storage for approximately 100 hours of HD content or 260 hours of standard-definition programmes. Multi-media compatibility is limited, but you can view JPEGs and listen to MP3s stored on a NAS drive, connected to your network. One nice trick is the Humax Live TV app, which lets you push whatever you're watching on your mobile device to the TV.

Televisions with Freeview Play

Panasonic was the first TV manufacturer to launch products compatible with Freeview Play connected services.

Panasonic's 2015 Viera TVs will be updated automatically. Top of the range is the 65 inch screen Viera TX-65CR730B from £1800. At the lower-end is the 50 inch screen Viera TX-50CX680B, £650 from Tesco Direct, this model is also available with a 40 inch screen. In September 2015 the 50 inch Viera TX-CX802 (£1169 from John Lewis) was rated as a Best Buy by Which. All models are 4K Ultra HD with built in WiFi.

Panasonic will also introduce Freeview Play products including three Blu-ray recorders and two digital TV recorders. More products are in the pipeline from Humax, Manhattan TV and Vestel, and other major TV manufacturers are likely to follow.

Freeview Play vs YouView

Freeview Play outwardly looks very similar in what it offers to YouView, but it will not have the pay channels associated with the YouView packages available from BT, TalkTalk and Plusnet– and you won't be tied in to a provider in order to keep the service.

Differences to YouView do exist. One superior feature of Freeview Play on the Humax FVP-4000T is that it enables content to be streamed, via DLNA, to other devices on the same network. Another unique feature is that with three HD tuners onboard you can record four shows at once whilst watching a recording, or a live channel. Another option would be to say watch EastEnders live on BBC One in the living room, record Channel 4 News at the same time, whilst streaming Emmerdale to the kids watching on a tablet in the kitchen. Once any launch bugs and teething problems are ironed out, Freeview Play will offer a very attractive service.

Television Viewer's Guide

YouView

YouView offers Freeview television channels plus additional on-demand and catch-up content, such as BBC iPlayer, delivered over your broadband connection.

YouView provides access to Freeview TV channels, catch-up and on-demand services via a set-top box. You can purchase a YouView set-top box for a one-off payment. However if you are not interested in paid-for content, or phone and broadband services from the main YouView suppliers, we'd recommend you look at Freeview Play instead. It offers very similar TV functionality to YouView, but in a simpler package and you won't be tied in to a YouView provider to keep the service.

YouView providers
You can obtain a subsidised or free YouView box from the main YouView providers – BT TV, TalkTalk TV and Plusnet.

As well as the set-top box, you will need to connect the box to your TV aerial and will need a broadband internet connection offering a connection speed of at least 3Mbps to watch online content - faster if you want to take advantage of BT's 4K content.

What channels are available?
The range of mainstream channels will depend on Freeview availability in your area. If you live where a full range of services is provided then you should get around 60 TV channels, up to 12 available in high-definition and 25 radio stations. TV includes the main BBC, ITV, Channel 4 and Channel 5 services, plus Yesterday, Dave, PickTV and various music and shopping channels. Radio includes all the BBC national networks, and commercial services.

Pay content such as Sky Movies, Sky Sports and BT Sport is available, but exactly what's available depends on which company you choose for the YouView package. All three providers offer Netflix from £5.99 a month.

Are you covered?
Freeview provides reception of the mainstream TV channels, and requires an aerial, so if you have good Freeview reception it will be fine for YouView.

Broadband requirement
To receive YouView's on-demand and catch-up coverage you will need to connect your YouView set-top box to a broadband router. YouView recommends a download speed of 3Mbps, but 6.5Mbps is recommended for sport to stream the content smoothly.
You can check your broadband speed at the following website:
http://speedtest.btwholesale.com/

Does YouView work with WiFi?
Yes. WiFi connectivity is not built-in to the Humax DTR-T2000, nor the Huawei box used by TalkTalk, but you can use an Ethernet connection, or a HomePlug internet-over-mains system to get a connection from a remote router.

Is it best to buy outright or through a YouView provider?
It is possible to get a subsidised YouView box if you sign up to one of the company's packages. All offer the standard YouView service with their own premium upgrades. BT claims to offer the largest YouView video-on-demand library with 12,000 shows available including the UK and US TV shows, Blockbuster (pay per view) and classic films, children's shows and music videos. TalkTalk offers over 4,000 titles, and has options to obtain Sky Sports and Sky Movies content.

There are pros and cons to either buying a YouView box outright, or obtaining one as part of a package. Over time, outright purchase is likely to work out the least expensive option and this will be even more the case as additional models come on to the market and prices fall. On the other hand, with a package from BT, TalkTalk or Plusnet you should receive additional technical support and access to extra programme content, albeit at a higher long-term cost.

YOUVIEW 23

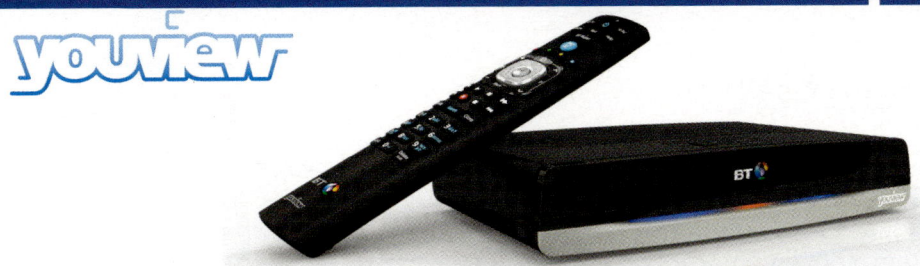

Humax DTR-2110, supplied with new BT TV packages.

Factor in the broadband costs
A very important factor to consider is the cost of broadband programme viewing or downloads. With a box from BT, TalkTalk or Plusnet check what broadband allowance is provided. If you purchase a YouView box outright it is important to check that your broadband allowance will cover your on-demand and catch-up viewing. Downloading or watching video uses large amounts of bandwidth and can be expensive.

Recording TV
Fitted with twin tuners, YouView+ set-top boxes work as Freeview PVRs giving the ability to watch one programme and record another at the same time in either standard or high-definition. Alternatively, two Freeview channels can be recorded simultaneously. Live TV can be paused and rewound including in smooth slow motion, ideal for scrolling back through credits etc. Series recording and smart recording to resolve overlap clashes are also provided. Storage of up to 300 hours of SD or 125 hours of HD content is provided by the internal 500GB hard drive. HDMI, SCART and component audio and video, plus S/PDIF for external audio, is provided, as is a USB connector which, in future, might be used for connecting to additional storage devices.

Electronic Programme Guide
A useful feature of the YouView system is its two-week EPG, which goes back 7 days as well as the standard 7 days forward from real time. The scroll-back guide pulls together all available catch-up TV content from across the BBC iPlayer, ITV Player, All 4 and Demand 5. The EPG is fully searchable, simply by typing in the first few letters of the desired programme name via the remote control.

Recommended equipment
You don't need a new TV as YouView functions are built-in to the set-top box, but an HD TV is ideal as YouView supports full HD. On-demand and catch-up programmes are also available in HD. In 2015 Sony became the first TV manufacturer to offer built-in YouView through its 2015 Bravia range, giving users access to YouView without the need for a separate set-top box.

Humax DTR-T2110, £150, is the same as the box supplied by BT. It offers reasonable performance. BT's latest box is the Humax 1TB Ultra HD YouView+ set-top box and will be needed if you want to take advantage of BT's 4K content.

TalkTalk's Huawei manufactured YouView box only has a 320GB hard disk, compared to the standard 500GB installed in the Humax DTR-T2100.

Catch-up
Internet connectivity is required for catch-up TV services: including BBC iPlayer, ITV Player, All 4, Demand 5, STV Player, Milkshake, UKTV and S4C. This is arguably YouView's best feature. It enables virtually any material broadcast over the previous week to be watched. Some content, for example films on Film4, is not available due to copyright issues.

Setting it up
Setting up a YouView box is relatively straightforward. You'll just need to connect your terrestrial television aerial input, an HDMI or SCART connection to your television and an Ethernet cable to your router.

Television Viewer's Guide

Freesat

Not to be confused with FreesatfromSky, Freesat is a subscription-free satellite system: conveniently you can use a Sky dish to pick it up.

Freesat involves no contract or subscription, just the cost of the box (which may be a set-top box or built-in to a television), a dish and their installation.

The key attractions of Freesat are that it can offer digital TV coverage in areas where Freeview terrestrial coverage is poor, and it offers a way to pick up subscription-free high-definition television. The launch of Freetime has added a catch-up capability and on-demand functionality too, and mirrors what's happening with Freeview with Freeview Play and YouView.

Confusingly, the name Freesat is very similar to the name 'FreesatfromSky', a competing service from Sky.

What equipment do you need?

The basic equipment needed for Freesat is a satellite dish and a set-top box, or a Freesat-equipped TV. If you're planning to use a Freesat PVR, with two tuners, you'll need two feeds from the LNB on your satellite dish, just as you do with Sky+. For high-definition (HD) viewing you'll need an HD television.

Using a Sky dish

Because they use the same satellite group (and even the same transmissions, working under a different Electronic Programme Guide) you can use a Sky dish. For those on communal aerial systems in flats, it should only be necessary to buy and connect a Freesat box to get the full service.

If you don't have a satellite dish it will cost around £80 for a simple standard dish installation.

Do you need a new TV?

No. As with YouView and other platforms such as Sky, most users will use a set-top box that provides the Freesat functionality. There are television sets with a built-in Freesat tuner, useful if you want to keep your television set-up clutter free, but not so flexible if you want to upgrade your Freesat equipment.

What channels are available?

Freesat carries around 180 TV and radio channels. They're listed in the guide. Although Freesat carries most of the popular free channels that are on Freeview and Sky, there are a few channels such as Dave that are available on Freeview but not on Freesat.

Non-Freesat EPG satellite services

Some Freesat boxes such as the Humax HDR-1010S can also operate as a standard satellite receiver in 'STB' mode, enabling them to be used for receiving free-to-air channels not available on Freesat's EPG: though functionality is limited compared to the Freesat-listed channels.

HD channels

The main BBC HD channels, as well as ITV HD and Channel 4 HD are available subscription free.

Freetime

In September 2012 Freesat launched a new onscreen TV guide called Freetime. The service offers new features based around a simple and easy-to-use programme guide.
- A TV guide letting you turn back time and watch programmes that have already been on, but you may have missed.
- Now and Next view to show what's on.
- A Showcase section that offers recommendations on programmes to watch tonight, in the coming week and on-demand.

FREESAT 25

The Freetime feature, offers a single programme guide giving easy access to the coming, and past week's programmes as well as on-demand and Showcase content.

• On-demand TV from BBC iPlayer, ITV Player, All 4, Demand 5 and YouTube.
• Improved recordings that make it even easier to record TV at the touch of a button, and find episodes automatically filed as series.

Broadband requirements
To to make use of Freetime with its catch-up content you will need to connect your set-top box to a broadband router. Freesat recommends a download speed of 2Mbps, but 5Mbps is better to ensure smooth streaming.

WiFi
Most boxes have an Ethernet connection, and the latest boxes usually have built-in WiFi. If your set-top box does not support WiFi it is possible to use a HomePlug internet-over-mains system or Humax's optional wireless dongle.

The Freetime EPG
The Freetime EPG enables you to look back over the last week to find past programmes. The 'Showcase' menu features highlights of what's on catch-up or live TV and can be used to set recordings. Radio programme details are also carried. These can be used to schedule recordings with a Freesat PVR.

Are you covered?
Freesat offers similar coverage to that of Sky. Unless you have trees, hills, or tall buildings in the way you should be able to get a signal anywhere in the UK.

Recording Freesat+
As with other platforms, the '+' sign is used to indicate a recording capability. Freesat+ boxes can pause, record and rewind in the same way as with recording equipment on the Sky or Freeview platforms.

Set-top boxes
Freesat seems to be currently pushing three set-top boxes, you'll also also find the Manhattan Plaza HD-S2 (£44), available from retailers such as Currys, and Bush models from Argos.

HUMAX HB-1000S £79
This offers Freetime, but lacks a disk for recording, although it will take an optional USB-linked disk. There is only a single tuner so you won't be able to watch one channel while you record another. WiFi is not built-in.

HUMAX HDR-1100S £189
The HDR-1100S gives access to Freetime and catch-up services such as BBC iPlayer, ITV Player, All 4, Demand 5 and YouTube, plus movies from Curzon Home Cinema. The box offers twin tuners, WiFi, and capacity for around 250 hours of SD recordings on the 500GB hard disk. It is available in glossy black or white finishes.

HUMAX HDR-1100S (1TB) £220
This 1TB box has all the same features as the standard HDR-1100S, but the 1TB hard disk provides twice the recording capacity.

HUMAX HDR-1100S (2TB) £299
As the model above, but with a 2TB hard disk for additional storage.

Freesat integrated televisions
Integrated Freesat TV sets are available, primarily from Panasonic and Samsung. Freesat's website has a good list of products, with prices ranging from £749 for a 40 inch model, to £2700 for the 65 inch model TX65CX802B 4K Ultra HD LED screen.

We wouldn't recommend upgrading your TV set just for the benefits of having Freesat built-in. Instead we would recommend you simply purchase a set-top box with the required level of Freesat functionality.

Television Viewer's Guide

Sky

Sky offers an excellent range of standard and high-definition programming via satellite. If you don't wish to pay a monthly subscription the 'FreesatfromSky' package offers installation of a dish and a basic Sky digibox for £175. Sky's NOW TV device is a useful way to get Sky and other content via broadband if you don't have a smart TV.

Cut-price installations, a good range of channels with some of the best movies and sports programming, and the excellent Sky+ PVR have all helped Sky become a dominant force in UK broadcasting.

NOW TV
NOW TV is Sky's internet service. It carries Sky's premium movie and sports programming, as well as content from services such as BBC iPlayer, ITV player and YouTube. With a NOW TV device you can pay for daily, weekly or monthly passes to access Sky's premium content.

Channels available
Sky carries a wide range of UK TV and radio channels. Many are free, others require subscription. You will have to pay for the Family Bundle to receive the 3D content. Listing magazines such as *Radio Times* show the programmes carried by the main channels on Sky.

Free channels
You do not need to pay a monthly subscription if all you want are free channels. These are colour-coded in our channel list.

FreesatfromSky
This scheme launched in 2004 and should not be confused with the Freesat satellite service provided by the BBC and ITV. Viewers who take up Sky's offer receive a satellite dish, digibox, viewing card and installation, all for a one-off payment of £175. You'll get the new SkyHD box to watch FreesatfromSky – and it's yours to keep. There's no annual contract and it's easy to upgrade to Sky's pay packages if you are so inclined. The disadvantage of this package is that you do not get Sky+ recording for free. For this you will have to subscribe to a programme bundle.

Equipment and installation costs
Sky offers various deals and packages cutting the cost of the digibox (usually £49) and offering low cost installation (usually £10) depending on which TV package you subscribe to. Look out for special offers where the subscription price is reduced or includes other incentives to join Sky.

The box currently supplied is the Sky+HD box. A larger capacity, 2 terabyte, box is available with most bundles for an additional cost. It provides around six times more personal storage for recording. Boxes now support WiFi, so you can connect by WiFi to a home broadband router rather than running cables round the house. The 2TB box provides space for up to 1180 hours of SD, or 350 hours of high-definition recordings. Customers with older equipment can usually upgrade to the new box.

Installations sometimes incur a set-up cost of between £30 and £60, and you will be required to sign a 12 month contract during which time the digibox should be connected to a telephone line.

Sky+ recording system
Sky+ provides an excellent way of recording TV and radio programmes from Sky. There's further information about Sky+ on page 28.

Use your phone as a remote control
The Sky+ app lets you use your iPad, iPhone or iPod Touch as a remote control. It's compatible with the latest Sky+HD boxes. You'll need an active WiFi connection and the device needs to be operating at least iOS 5.

With the Sky+ app on your Android tablet or mobile, you can set your Sky+ box to record a show or series that's coming in next seven days, look through your favourite channels, search for shows and check out recommendations. If you have a Sky+HD box, you can also use your Android device as a remote control, and browse your Planner.

Sky High-Definition
If you don't subscribe to a Sky programme package, and have FreesatfromSky, you will receive the public service HD channels. This

SKY 27

is now on a par with Freesat and Freeview which both offer free access to HD.

4K on Sky
In November Sky announced a new range of products including the 'Sky Q Silver' set-top box. Sky Q products will offer a blend of live and on-demand TV including 4K Ultra HD content. These will launch in early 2016 along with a 4K Ultra HD service carrying sport, movies and entertainment content.

Remote Record
If you have a Sky+, or Sky+HD, subscription you can remotely program your Sky box to record using a smartphone running Sky's app or on a computer via the Sky website. The service is currently free, but SMS text requests cost 25p each.

Subscription packages/bundles
Sky sells packages of channels in what it now calls bundles - the Original Bundle, Variety Bundle and Family Bundle. See page 29 for further information.

3D TV
Limited 3D content is available via the Family Bundle's on-demand programming.

Sky Catch Up TV
Sky has renamed its on-demand services 'Catch Up TV'. All TV bundles come with Catch Up. You'll need to connect a WiFi-compatible black Sky+HD box to your broadband router to access it. With older boxes you may need a WiFi connector from Sky. If your Sky+ box is close to your router you can connect with an Ethernet cable. The service offers catch-up TV services from Sky channels, ITV Player, Demand 5, BBC iPlayer and All 4, TV Box Sets and Sky movies. The content available depends on your subscription.

Catch Up makes use of partitioned space reserved on the most recent Sky+ boxes.

When the Sky box is not busy the service sends programmes to a reserved portion of the box's hard disk.

Sky Go
Sky Go is available free to existing Sky customers and enables you to watch Sky channels and programmes on PCs, laptops, mobiles and tablets at no extra charge. If you are not a Sky customer, subscriptions are available from £20 per month.

Overseas/foreign channels on Sky
Sky has very limited overseas offerings. Euronews is available in six languages, including French, German and Spanish.

Subtitles - audio description
To access the Subtitles and Audio Description menus press the Services button on your remote control, select Sky+ Setup; following this you can select subtitle and audio description settings. Don't forget to select Save Settings when you've finished.

Using Sky overseas
Using a Sky digibox outside the UK is a breach of contract and Sky will switch off your card if it finds out. However many people do this. You will find more information about using Sky abroad, including details on how to set up a Sky dish on page 106.

Sky Pros
Sky has a good range of TV and radio channels, and an excellent EPG. Sports coverage is good too. Sky+ offers superb recording features. Installation is often provided free of charge depending on the packages being taken.

Sky Cons
The cost of subscribing to programme packages and the requirement to pay for Sky+ recording. Some people may find obstructions such as trees or nearby buildings prevent the easy installation of a satellite dish.

Television Viewer's Guide

Sky launches Sky Q

Sky announces Sky Q – a range of products including the 'Sky Q Silver' set-top box that will offer a blend of live and on-demand TV. The 'Silver' box is 4K ready and offers streaming to other devices in your home.

In mid-November Sky announced
a new range of products including the 'Sky Q Silver' set-top box. Sky Q products will offer a blend of live and on-demand TV including 4K Ultra HD content. The new set-top boxes will launch in early 2016, but prices have not yet been announced. This is the most significant new product launch for Sky since it rolled out its HD packages a decade ago. The new service looks as if it will be offered separately to Sky's existing satellite offering and its NOW TV on-demand service.

New set-top boxes
Dubbed Sky Q Silver, the 2TB set-top box comes with a whopping 12 channel tuners. These will let you record up to four programmes at a time while watching a fifth, and also send signals to Sky Q Mini streamers with live TV. These can be used to pickup Sky content sent to other TVs you may have around the house – in a similar way to how Sky multiroom works now. The set-top box will also support streams to two tablets using a new Sky Q app.

The box comes with 2TB of built-in storage, letting you record roughly 350 hours of HDTV. We do know that the premium Sky Q Silver box is 4K Ultra HD ready, but we don't yet know whether it'll include support things like HDR and 5.1.4 surround sound.

The company is also producing a barer-bones Sky Q non-4K box that comes with 1TB of storage and the ability to support fewer external devices, but is otherwise the same. The new boxes will come with a new Bluetooth-enabled remote control that has a touchpad and built-in voice controls.

Is a new dish needed?
One thing that current Sky TV customers will want to ask is 'Do I need a new satellite dish to get Sky Q?' The answer to that is no, the same dish will be fine. However, you will need a new LNB fitted. Sky's engineers will upgrade this during the installation process.

Sky Q Mini streamers
In the past, if you've wanted to get Sky TV in more than one room, it's been an expensive proposition. Sky Q Mini promises to change that; the mini streamers have more in common with Now TV boxes than they do with a separate Sky+HD box.

The idea is that the Sky Q Silver box will sit in your living room and you'll set up Q Mini streamers in up to two other rooms. These connect to the TV via HDMI and to your home network over WiFi or via a Powerline adapter, removing the need to have extra satellite cables running up and down your home. You can get access to the same content as you would on the Sky Q Silver box, although apparently not any Ultra HD content – the spec sheet lists 1080p as their maximum supported resolution.

Sky Q Hub
Usefully, the Sky Q Silver box also doubles as a WiFi hotspot, and this will enable you to sync and download content to tablet devices. There is a catch – this feature is only activated if you've got a Sky Q Hub and to get a Sky Q Hub, you'll need Sky Broadband.

The Sky Q Hub ties everything together. Not only is this an ADSL2+ and VDSL compatible router, it's designed to communicate with the Sky Q Silver and Q Mini set-top boxes, handling programme requests and sending content to the boxes over WiFi and Powerline.

Content
Sky will launch an Ultra High Definition service in 2016. The new service will offer customers a range of sports, movies, and entertainment content with up to four times the detail of HD. This may not be ready by the time the Sky Q Silver launches – and while the spec sheet mentions 2160p as a supported resolution, it also mentions that it'll be added with a future update.
www.sky.com/shop/tv/sky-q/overview

SKY+ AND SKY CATCH-UP

Recording on Sky

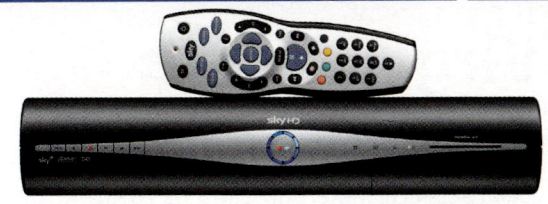

Sky+HD set-top box

Sky+ offers a simple and effective way to record television and radio. With the Sky+HD two terabyte box you can store up to 350 hours of HD recordings.

Sky+ launched in September 2001. The system uses a set-top box with twin tuners, enabling you to watch one programme while simultaneously recording another. Alternatively you can record two programmes while watching a recording already made. All Sky+ boxes can record radio channels and the latest ones will give you access to Sky's catch-up services if you connect your box to your broadband router. These offer content from services such as BBC iPlayer.

For many years Sky+ offered the best way to record programmes but now services such as Freeview Play, YouView, Freesat and Virgin media offer similar features, some without the need for a monthly subscription.

Simple to set recording
Setting up recordings and finding programmes to watch is very easy, in no small part due to the clearly laid out electronic programme guide (EPG). You can pause live TV, rewind back through programmes you have been watching, and set up series recordings very easily. The latest search options make it easier to find programmes by title, actor, sport or any other keyword.

Set-top box specifications
New Sky boxes come with WiFi as standard. The basic Sky+HD box is supplied free with most packages, and offers up to 60 hours for HD recording or up to 185 hours in standard definition. The Sky+HD 2TB box provides 1.5TB for personal storage, giving you a recording capacity of up to 350 hours in HD or up to 1180 hours in standard definition TV. You also get 500GB for a selection of catch-up TV that Sky automatically downloads to the box.

Box upgrades
Existing Family bundle customers can upgrade to the Sky+HD 2TB box from £49, plus an optional £60 set-up fee. If you would like to upgrade we'd suggest that you haggle with Sky to get the best deal! If you've an older box and don't want to pay for an upgrade, Sky sells a WiFi connector device. It will enable you to connect compatible Sky+ boxes to your WiFi router wirelessly, giving access to Sky's catch-up content.

Sky Catch Up
This makes use of the partitioned space reserved on your Sky+ box. The programmes available depend on your TV package. Catch-up TV from BBC iPlayer, ITV Player, All 4, 5 On Demand, plus a selection of Sky channels is also available. Sky movies are available for rental online via the Sky Store. Downloads usually count towards your broadband usage allowance.

Sky+ remote recording app
With the Sky+ app you can send instructions to your Sky+ box to set up a recording while you are away from home. You'll need a compatible smartphone, laptop or tablet device. The app also enables you to use compatible Apple or Android smartphones and tablet devices as a remote control. It's compatible with the latest Sky+HD boxes, but you'll need an active WiFi connection to use it.

Sky+ costs
New customers receive a free Sky+HD box when they subscribe to a Sky programme bundle. If you don't want to pay monthly for a programme bundle Sky charges £10 per month for the recording features of Sky+.

Television Viewer's Guide

Virgin Media

Virgin Media provides television, broadband, landline and mobile phone services to UK customers via its cable network.

The first thing to say about Virgin Media is that only around 50% of the population have access to its cable service. If you are not covered you will need to look for an alternative platform, such as Freeview, Freesat or Sky.

Virgin set-top boxes
Virgin supplies two types of box. The TiVo set-top in 500GB and 1TB configurations and the older style 160GB VHD box. The VHD box is a cheaper option but lacks the recording capability of the TiVo. New boxes also have the benefit of an enhanced 14 day EPG with integrated catch-up content.

TiVo
Virgin charges £5 monthly rental for TiVo boxes; to upgrade from the 500GB to the 1TB TiVo box will cost an additional £50. If the box fails due to a fault, Virgin is responsible for its repair or replacement.

TiVos have three tuners which can make simultaneous recordings. All of them have their own buffers, so you can be recording two programmes, watching another, and still flick between the three and browse back through the buffered video as you wish. By comparison the Sky+ HD box only has a buffer on the currently watched channel - so you can't change channels and still rewind.

What channels are available?
If you are looking for a free or cheap way to pick up TV, Virgin Media is probably not for you, unless you want to take out Virgin's Big Easy bundle providing broadband and phone line rental too. If you are prepared to pay for subscription TV, Virgin offers a comparable service to Sky's, with TV bundled into packages. For a full list of what is available see: www.virginmedia.com

What about HD and 4K
Virgin includes over 10 HD channels with its packages; and 43 HD channels with its XL package. The basic channels include BBC One HD, BBC Two HD, Channel 4 HD, Channel 5 HD, Film4 HD and ITV HD. A fibre-optic cable is required to receive Virgin's forthcoming 4K UHD programmes.

3D TV
With Virgin's M+, L and XL TV packages, you'll get a certain amount of 3D content to watch for no extra cost. You can also order 3D movies on-demand at any time from £5.99.

Sky Sports and Movies
Virgin customers can subscribe to Sky's Sports and Movies packages. Subscription to both, for example, costs £45 per month on

The Virgin TiVo set-top box offers good functionality including three tuners.

VIRGIN MEDIA - CABLE TV

mid-tier programme packages, but is included within Virgin's Big Daddy bundle. With most Virgin TV packages you can pay for additional Sport, Movie and Adult content.

Recording TV
Virgin's TiVo boxes offer both HD and standard-definition recording. The 1TB TiVo box, for example, gives around 536 hours of SD or 121 hours of HD recording capacity.

On-demand and catch-up
Catch-up TV from BBC iPlayer, ITV Player, All 4 and Demand Five is included with all the TV packages.

All chargeable on-demand content must be viewed within 48 hours.

Virgin TV Anywhere
This gives subscribers online access to many of Virgin's channels on iPhone, iPad, PC or Mac alongside several thousand hours of on-demand content.

All you need to use Virgin TV Anywhere online is a desktop or laptop with a web browser. TiVo customers can also download the free Virgin TV Anywhere app for iOS, Android, Amazon Kindle Fire and Fire phone. The service can be used on a maximum of two devices and is only available for use in the UK.

The app will also allow Virgin Media TiVo customers to set up, manage or delete recordings and to use your device like a remote control.

Broadband considerations
One of the advantages of a Virgin package with broadband is that on-demand and catch-up viewing doesn't count towards your broadband allowance. With its fibre optic cable Virgin Media offers some of the fastest broadband services in the UK. However many factors affect the broadband speeds you receive, and there are big differences between Virgin and Sky, for example, on this front. Virgin Media broadband has come in for some stick for its traffic management policy which slows heavy users during peak times, and this is something worth bearing in mind especially on their cheaper packages. Packages are advertised at 'Up To' speeds and these will not always be achieved.

VIRGIN PACKAGES

Virgin offers a range of packages, some with TV, phone and broadband bundled together. A TiVo 500GB HD set-top box is included. Activation is free on the 500GB box, but a rental fee of £5 per month is charged. Activiation costs £50 for the 1TB set-top box.

Virgin Media frequently offers half price discounts for the first 9 months. Packages require a 12 to 18-month contract.

TV PACKAGES

TV M: 60+ TV channels, 10 HD channels and catch-up players – such as BBC iPlayer. The basic TV M package is only available as part of Virgin's **Big Easy** bundle. This costs £22 per month, plus £50 installation fee, and £17 monthly Virgin phone line rental. If you pay for the line rental up front you can make a saving.

In addition to the TV content you'll get up to 50Mbps broadband and unlimited weekend calls to UK landlines and Virgin Mobile numbers.

More TV: **£18** per month. 70+ TV, 10HD, 8 Sky channels and 4 catch-up players.

M+: **£21** per month. 130+ TV, 11HD, 10 Sky channels plus Virgin Anywhere, TV on-demand and catch-up services.

Size L: **£28.50** per month. 170+ TV, 11HD, 10 Sky channels plus Virgin Anywhere, on-demand and catch-up services.

Size XL: **£36** per month. 230+ TV, 43HD, 14 Sky channels plus Virgin Anywhere, TV on-demand and catch-up services, BT Sport and ESPN.

BUNDLES

Virgin also sells what it calls 'Bundles'. These include a mix of broadband, TV and phone services. The Big Easy bundle is described above. At the other end of the scale the Big Daddy bundle costs £104 per month, plus £17 monthly line rental. It provides 260+ channels, Sky and BT Sports, Sky Movies, plus catch-up and on-demand TV.

Television Viewer's Guide

BT TV

BT TV offers a hybrid system with a mix of Freeview terrestrial TV and content delivered via the internet. As well as offering Freeview channels you'll gain access to YouView catch-up and on-demand content.

BT TV offers a mix of Freeview and internet-delivered content in a YouView-based system. You'll need an aerial and a fast broadband connection. BT TV is only available to BT broadband customers so you'll only be able to access it if BT already supplies your internet or you're happy to switch providers.

New 4K sevices launched in 2015
BT was the first UK broadcaster to have an Ultra HD 4K channel. BT's new Ultra HD TV 4K channel launched on August 2nd with the FA Community Shield match between Chelsea and Arsenal. It uses a Humax 1TB Ultra HD YouView+ set-top box and costs £15 a month for a package that includes its new Ultra HD channel. It will only be available to customers on its Infinity 2 product, with broadband line speeds of no less than 44Mbps. A £44 intallation charge is required for the 1TB Ultra HD Youview receiver. However, access to Netflix and its UHD offering weren't available at launch. BT says that it is working on making it available soon.

BT TV is YouView based
BT uses a YouView-based system to deliver its TV content. Essentially you receive your 'mainstream' TV channels via Freeview, with additional internet TV such as catch-up TV delivered by YouView. On top of this BT adds extra content, such as BT Sport, to make the whole package more attractive to its subscribers.

What do you need?
You'll need Freeview coverage, a BT landline and a BT Broadband subscription, and a BT YouView set-top box.

Set-top boxes
BT has updated its YouView box to the BT-badged YouView+ box (Humax DTR-T2110). It offers exactly the same features as the previous Humax DTR-T2100 box. The 500GB hard drive will record around 300 hours of content in standard definition or 125 hours in HD.

Equipment set-up
It is relatively simple to install a BT TV box yourself. As well as the set-top box, BT provides a BT Home Hub wireless router, and a 5 metre cable to connect to your router. You could use Powerline plug adapters if your TV is in a different room from the BT Home Hub.

BT TV packages
BT offers various packages. Some of the main ones are detailed below.

TV Essential + broadband
Gives up to 80 Freeview channels, access to iPlayer and a YouView+ box. Costs are £5 per month, plus a standard line rental of £18 per month, plus a £49 activation fee. You'll need to sign-up for a 12 month contract. This package provides 10GB broadband allowance, free weekend calls and free access to the BT Cloud and BT Sport. Unlimited broadband costs a further £6 per month.

TV Entertainment + unlimited BT Infinity
All the features of the Essentials TV pack, plus an extra 28 channels. You'll also have much faster broadband, but the fibre-optic BT Infinity service is only available in limited parts of the country. It costs £35 per month, plus line rental of £18 per month. The activation fee has been dropped.

BT Sport
BT Sport is available free of charge to BT broadband customers with BT Infinity fibre-optic broadband or fast broadband.

Sky Sports and Movies
Sky Sports costs an additional £22-£27 per month - you will need the older Vision+ set-top box and ideally have BT Infinity broadband. Sky Movies costs £13.50 per month.

Netflix
This is available at £6 per month.

Television Viewer's Guide

TalkTalk TV

Very similar in proposition to BT TV, TalkTalk TV is another YouView-based system. You'll need Freeview coverage for the main channels and broadband from TalkTalk to deliver the extras.

Like BT TV, TalkTalk TV is a 'hybrid' system, in that it offers a mix of Freeview and broadband content. See the YouView section in the guide to see how it works.

What do you need for TalkTalk TV?
As the main channels are carried on Freeview, the first thing to do is to check that you are in a Freeview coverage area. You'll also need a phone line and TalkTalk broadband with a minimum line speed of 3Mbps, but 5Mbps is needed if you want to watch on-demand content such as Sky Sports.

Set-top boxes
TalkTalk has stopped issuing Humax equipment and now supplies Huawei boxes to new customers.

TalkTalk packages and charges
To receive TalkTalk's TV services you will need to subscribe to one of their packages. Two are available: Essentials TV and Plus TV. You will need to sign up to an 18 month contract to TalkTalk's broadband and phone line rental too. New lines may be subject to a £20 connection fee.

Essentials TV
TalkTalk's basic package provides the non-recording Huawei DN360T. You will be able to pause and rewind live TV, and access the on-demand players BBC iPlayer, ITV Player, All 4 and Demand Five, Now TV, Milkshake, UKTV, S4C, TalkTalk and Sky Store. The cost is £10 per month (offers start at £7.50 per month) for unlimited broadband, plus £17.70 line rental. You'll also receive free evening and weekend calls to UK landlines.

Plus TV
With this package you'll get a Huawei DN3710T set-top box with twin tuners, which offers hard disk recording (currently only of broadcast channels, but soon to include broadband channels) as well as a more generous phone tariff, with unlimited anytime calls to UK landlines.

At the time of writing TalkTalk were offering Plus TV for £10 per month for the first 6 months, rising to £20 thereafter. In addition there's a £17.70 monthly line rental. You also get six Sky Channels from the Entertainment boost, and like the Essentials package you get unlimited broadband. If you opt for one of Talk Talk's faster 'Fibre' broadband options you'll pay an additional £7.50 per month.

TV Boosts
These are available for an additional monthly cost. Individual boosts range from £5 to £40 a month. Boosts include entertainment channels, children's channels, Picturebox, Sky Sports, Sky Movies, box sets, world TV such as Brazil, Asia and Greece and music. An advantage of TalkTalk's Boosts is that you only need to sign up for a month at a time.

Broadband allowances
One of TalkTalk's claims is that their broadband is totally unlimited and that they will not restrict the speed of your connection.

Which is better TalkTalk, or BT TV?
BT has mounted a challenge to Sky, pursuing higher income pay-TV households with its heavy investment in BT Sport and live Premier League football. BT has also been a market leader with its introduction of a 4K channel. TalkTalk on the other hand has targeted more cost-conscious households with cut price deals and the option of a cheaper set-top box. The Essentials package halves the monthly cost of TalkTalk's internet-based television service.

The decision as to which service to choose is not always simple, and you will need to consider the full range of what their packages offer, rather than just the TV content.

EE TV

In 2014 mobile operator EE started selling its home TV service. It's based round a smart set-top box that connects to your TV via an HDMI cable. The box offers a mix of Freeview channels, with catch-up and on-demand content.

EE TV is Freeview based
As with YouView-based packages, the core service is based on Freeview, giving you over 70 channels to view, including 13 in HD. This is supplemented by additional catch-up and on-demand facilities such as BBC iPlayer.

Available only to EE customers
EE TV is free to eligible EE, Orange or T-Mobile customers; the bad news is that this means you have to subscribe to both EE's mobile and broadband services to qualify. If you are already on these services then EE TV looks an attractive option, if not you will need to switch. As usual, you also need to pay the fixed-line charge, and there is a minimum 18 month contract. The cheapest EE broadband package offers EE TV, weekend calls and 17Mbps internet for £12.95 per month. Fibre internet with weekend calls at 38Mbps is £22.95 and 76Mbps fibre and weekend calls is £32.95 per month. All include EE TV for free. All prices are not inclusive of line rental though, this'll set you back a further £16.40 per month.

Set-top box
A little larger than a standard router, the set-top box has dual band WiFi and an Ethernet socket. With a 1TB hard disk you can record around 600 hours of SD content, roughly half this for HD. It also has four internal Freeview tuners, increasing its flexibility.

Facilities
Unique is the ability to simultaneously watch 4 different channels on EE TV, one on your TV, three on tablets or phones connected by WiFi. This means arguments over who watches what should be a thing of the past - although it will place demands on your broadband service.

Another useful feature is the ability to use your phone or tablet as a sophisticated remote control for EE TV, with facilities over and above those of competitive services. Essentially it enables you to control the user interface from the phone or tablet, including live recording. Of most note is 'Flick', the ability to transfer what is on your tablet to the main TV with just a flick of your finger. Note these capabilities are not available to Windows Phone users, who have to rely on the supplied remote.

Replay and Restart
Also controllable from the remote is the ability to 'rewind' a live programme to the start and watch it uninterrupted. Called Restart, it isn't unique but is nevertheless useful. Replay is aimed more at on-demand programming, enabling you to constantly record up to six channels so that you can watch anything in the preceding 24 hours.

What content is available?
BBC iPlayer and You Tube are available, as well as Wuaki.tv, a Netflix type service. Demand 5 is also there, in addition to numerous more niche services. However EE TV does not offer Netflix itself, nor Amazon Instant Prime (Lovefilm as was). All 4 and ITV player are missing, reportedly due to concerns that EE Replay lets viewers fast-forward through adverts on hundreds of hours of TV shows instead of watching the ads inserted into their catch-up services.

What future for EE TV?
With BT's purchase of EE this year, there exists some uncertainty as to the future for EE TV – particularly as BT already has its own TV package, BT TV.

Amazon Fire TV

Amazon's updated Fire TV set-top box gives you access to a range of on-demand video content, music, photos, games, and 4K.

The Amazon Fire TV 4K, £80, is the second-generation model of the Fire TV set-top box. It supersedes last year's model, which is no longer available, and provides 4K Ultra HD video streaming.

Content

You'll find a good range of music, movies, and TV shows through the Amazon Prime Instant subscription service, and you will also be able to access Netflix, Sky News and BBC iPlayer through the apps section. There are a few missing however, such as ITV and Channel 4 catch-up services.

The key feature, however, for the new Fire TV box is the dedicated section for 4K Ultra HD video. This enables you to purchase or rent a selection of 4K content to play through the box to a compatible TV. However 4K content from both Amazon and Netflix is limited at the moment. Many popular TV shows and movies aren't yet available. At present, it's mainly the Amazon and Netflix original series that are in 4K.

Fire TV lets you stream a wide range of TV programmes, films, and other videos from Amazon's vast library of content. Free material is available; you can also rent or buy individual films or shows.

Costs

The box itself costs £80. The underlying service is really geared to Amazon Instant Video customers though. This is available for £6 per month or as part of Amazon Prime membership, £79 pa. Prime subscribers can also stream 4K versions of Amazon's original TV series at no extra cost.

Connections

If you are looking to give an older 4K TV model an upgrade, make sure that your TV or monitor has HDCP 2.2 support and at least one HDMI 2.0 port. Although the latter probably won't be an issue, HDCP 2.2 – a copy protection for 4K content – isn't often found on older models of 4K TV.

On the back of the device you'll find an HDMI port for plugging into your TV, a power port for the included power adapter – an Ethernet port, a USB port and a microSD card slot for adding more storage. Although the box offers WiFi, connection via Ethernet is best, particularly for 4K video, and requires an internet connection of at least 15Mbps.

Photos and music

Photos you've stored in the Amazon Cloud Drive can be viewed on your TV. MP3s bought and downloaded from Amazon can also be played, as can any CDs you've bought that have AutoRip (whereby Amazon keep a copy in their Cloud). Finally you can upload your own music to the Cloud Drive to play through Fire TV, but this is limited to 250 tracks; something of a token gesture.

User interface and voice control

The user interface is simple and well thought out. Of most interest though is voice control. Using the microphone in the remote you can search for content with voice commands rather than entering characters one at a time. It works well, aided by powerful voice recognition facilities in Amazon's Cloud. The facility is limited to Amazon services though; it doesn't look across other services as some competitors such as Roku do.

A good service for Amazon users

Overall Amazon Fire TV is a slick service, with a powerful set-top box that looks good, and it delivers a wide range of content via a well designed user interface.

Internet Television

Internet TV options continue to expand. It's one of the fastest developing areas in UK broadcasting. Equipment ranges from Smart TVs, to set-top boxes, and to tiny devices such as the Amazon Fire TV stick. They all work in a similar way – and provide a link between your TV and the huge range of video content that is available on the internet.

The range of internet TV content is increasing rapidly. As well as being able to watch live TV from all the main UK broadcasters over the internet, you can catch up with programmes that you've missed, using services such as BBC iPlayer, for example. You can also watch on-demand content by downloading programmes or films to watch on your TV. New providers are entering the market too, and content is available from a wide range of sources; including Amazon, Netflix, Google and Apple. The TV world is certainly changing!

How is the content delivered?
For most viewers this is via a home broadband or better still, fibre-optic internet connection. You can also use a mobile phone data connection in conjunction with a phone or tablet device, but this can get expensive depending on your phone network charges.

Consider the broadband costs
A few words of caution, and this applies to any internet-delivered TV service. Be aware of the broadband usage charges. If you have a low-end package it may be that you have your usage capped, meaning you face additional charges or a reduction in speeds if you stream or download too much content.

Broadband usage rates vary, depending on the content and whether you are watching SD or HD. An hour of catch-up standard-definition TV viewing can consume as much as 800MB of your broadband allowance; HD up to 1.8GB per hour. This will increase substantially when 4K content is downloaded.

One thing worth considering is the selection of a broadband supplier that offers an unlimited/uncapped option at reasonable cost, or doesn't penalize you for watching online TV. You'll find that companies such as BT, TalkTalk, and Sky, for example, offer deals if you order your phone, internet, and TV from them. This can reduce the cost of internet TV, or enable you to watch without it affecting your broadband allowance.

Why would you want internet content on your TV?
The main reason is choice. You will be able to view an increased range of content. Services such as BBC iPlayer, ITV Player, All 4 and Demand 5 are widely available. They offer the option to watch live television, on-demand or catch-up, content that has been screened in previous weeks, and frequently a catalogue of stored programmes and films. The services carried varies widely between products and providers so check before you buy.

What equipment do you need?
One method to access internet TV is by using a Smart TV. However, as long as your TV is reasonably up-to-date we'd recommend that instead of investing in a new Smart TV, you simply buy a set-top box, or an adapter that has internet TV capability. Most of the latest set-top boxes from Freeview, Freesat, Sky, Virgin, TalkTalk and BT have a built-in internet capability. They all differ in the range of internet content that they offer, so check before you purchase.

As well as Smart TVs and set-top boxes from the main providers, there are many other devices, which enable you to connect your TV to the internet. Amongst these you'll find tiny devices such as the Amazon Fire TV stick and Google Chromecast dongle, that plug into an HDMI port on the back of your TV. Slightly larger, devices such as Sky's NOW TV box, Amazon's Fire TV, Roku 3 and the Apple TV box perform a similar job.

As already mentioned you will also need a broadband internet connection. TVs and set-top boxes usually connect to this with an

ONLINE / INTERNET TV

Ethernet cable, and this is our preferred option for speed and reliability, but WiFi connection is increasingly common.

Don't forget that you are not restricted to watching internet TV on your television. Many of the services available are designed so that you can watch on tablet devices and smartphones too.

Content from the UK broadcasters

Most of the main broadcasters offer options for viewing catch-up content or downloading.

BBC iPlayer – one of the best streaming and catch-up services - widely available with catch-up now available from the last 30 days.

ITV Player – offers streaming and a selection of shows from the last 30 days of programmes across five ITV channels, along with adverts.

Channel 4 – All 4 is Channel Four's internet service. It offers a variety of programmes recently shown on C4, E4 and More4, along with programmes from their archives.

Channel 5 – Demand 5 is the channel's internet service. Some shows are available to watch from the last 7 days, and others can be streamed from their archive.

Sky – Sky's On Demand has been renamed Catch Up TV. The service is available if you subscribe to a programme bundle. It provides catch-up from Sky channels, as well as from the other main broadcasters.

Sky Go – available to Sky subscribers, this enables you to watch Sky channels on PCs, laptops, mobiles and tablet devices.

What the main platforms offer

Freeview and Freeview Play Freeview HD products, both TVs and set-top boxes, have the capability to connect to the internet. You'll gain access to BBC iPlayer and other 'virtual' channels provided by Connect TV (delivered by transmissions company Arqiva). Freeview Play is Freeview's latest development. Similar to YouView, it offers subscription-free viewing and catch-up TV.

YouView provides a mix of Freeview channels delivered via your TV aerial, along with internet content. Catch-up TV content is available from BBC iPlayer, ITV Player, All 4 and Demand 5. YouView also offers a library of on-demand programmes, series, films and radio. Pay movies are available from NOW TV. If you connect your YouView box via BT TV, TalkTalk or Plusnet you will get access to an even wider range of content.

Freesat and Freetime

New Freesat set-top boxes, with the Freetime logo, have catch-up from the BBC, ITV, Channel 4 and Five. They also have the Freetime EPG, which lets you scroll back to find programmes available on catch-up services via BBC iPlayer, ITV Player, All 4, Demand 5 and YouTube.

Sky

Sky's Catch Up TV offers catch-up TV from the BBC, ITV, Channel 5 and All 4. In addition there are catch-up and box sets from the Sky-branded channels, films and UKTV channels such as Watch and Dave.
The latest Sky+ boxes come with built-in WiFi.

NOW TV is an internet service that offers Sky's premium Sport and Movie packages without having to sign-up to a subscription. You can also access content from BBC iPlayer and ITV Player. NOW TV can be accessed using devices such as Xbox, PlayStation, Google Chromecast and Apple TV. Sky also produces the NOW TV box. This costs £15, but you will have to pay to pick up Sky's premium Sport and Movie programming. You can do this on a monthly, weekly or daily basis. The NOW TV box is a good way to convert an older TV to give internet access.

Virgin

Virgin Media's TiVo set-top box is now supplied as standard with most packages. It offers a good range of online features including catch-up TV from BBC iPlayer, ITV Player, All 4, Demand 5. You'll also get apps giving access to Netflix, YouTube, Facebook and Twitter. TV series and films are also available on demand.

Apple TV

Apple's TV device undergoes an update and now offers Siri voice control and a new remote.

Apple TV device, from £97

In September Apple unveiled the fourth generation Apple TV device. With a range of changes the new box offers more processing power, extra memory and a new operating system, but Apple TV faces stiff competition from its main competitors: Google's Chromecast, Amazon Fire TV, Roku and Sky's NOW TV box. UK pricing hadn't been announced at the time of writing, but the 32GB model will cost around £97 and the 64GB version in the region of £130.

Specifications
The new box is much more powerful than its predecessor. It retains Ethernet and HDMI ports, and now supports WiFi and Bluetooth 4.0. The device runs on new software called tvOS. This has better links to Apple's mobile operating system, iOS, and allows developers to create apps for Apple TV – much like they do for the iPhone and iPad.

AirPlay is also supported so you can throw content or apps from your phone or tablet to the Apple TV.

The Apple TV app store is where you find apps you didn't know you necessarily wanted on your TV screen. It will feature games and other apps, but here Apple faces stiff competition from Google which is way ahead with the apps for its much cheaper Chromecast HDMI dongle.

Remote control
The new remote control has been widely praised. It has been given a makeover, and has a responsive glass touchpad at the top and volume rocker, menu, home and Siri buttons. It's light and is Bluetooth 4.0 enabled, needing charging every three months or so through the lightning port.
The remote comes with an accelerometer and gyroscope with gaming in mind. It's not perfect and the remote lacks a little precision that certain games require. There is the option to use a Bluetooth-enabled keyboard too.

Siri
Apple TV works with Siri voice control. This lets you find content from iTunes and Netflix using your voice – but not all apps will be supported – it is up to the app developers. When it is supported, using Siri can be easier than the process of keying in the programme or film you're looking for.

TV content
This is a little thin on the ground at present. Netflix is available – once you've downloaded the app, but no other major streaming services are, unless you're in the US where you can also get Hulu and HBO Now. There's no Amazon Video or UK specific apps like BBC iPlayer, All 4 and ITV. These may well come as developers get to grips with the new tvOS.

Delivering TV programming to viewers is widely expected to be the next step in Apple's entertainment strategy, but the company's much hinted at subscription TV service appears to have been delayed due to problems sorting out licensing deals for programming from content providers such as TV networks. Even when the service does launch, it is likely to be restricted to the United States

Who should buy?
On the positive side the box and remote are well made, the Siri remote is innovative and the tvOS comes with some useful features. On the downsides the box doesn't support 4K, so it's not future-proof.

At the moment we'd say the device is probably best for committed Apple users who want iTunes content on their TV. In the UK it may well struggle against cheaper rivals like Google's Chromecast and Amazon's 4K-capable Fire TV.

Television Viewer's Guide

NOW TV

Sky-owned NOW TV is an internet TV service. It is available to watch via computer, mobile devices, some game consoles and set-top boxes, and with a NOW TV branded set-top box.

NOW TV box, £14.99 including postage and packaging

NOW TV is designed for people who have no existing pay TV service. It gives access to Sky's subscription-based entertainment channels, sports and movies, on a daily, weekly or monthly basis, without having to sign up for a lengthy contract. It's a useful way to add internet TV if you don't already have a Smart TV or smart set-top box. In August Sky released a new version of the small set-top box.

Content

NOW TV is split into separate services including; Movies, Sports and Entertainment.

A Sky Movies Month Pass currently costs £9.99 a month for all of the Sky Movies channels available both streamed as they are shown, and on-demand.

A Sky Sports Day Pass currently costs £6.99 and provides access to seven Sky Sports channels, including Sky Sports F1 and the recently launched Sky Sports 5, for 24 hours. This is ideal if you want to watch a specific event and don't want a regular, contracted subscription. You can also pay £10.99 for a week or £31.99 for a month's access.

An Entertainment Month Pass costs £6.99. This provides access to 13 pay-TV channels not available on Freeview.

BBC iPlayer and YouTube

If you are using the NOW TV box you will also get access to additional content and apps. Functionality mirrors that of the Roku platform (the NOW box is a re-badged Roku device), with numerous apps available. BBC iPlayer gives access to live and catch-up TV and radio, useful if you don't already have it. Demand 5 is also available, and so is ITV Player. Various news channels can be watched, as well as specialist channels on Indian Food, Bollywood, and the like.

A Facebook app enables you to stream TV and photos from your friends, and Tune-In radio gives you internet radio. Spotify is also available, although only if you have a Premium licence for it. If you have an online photo managing account with Picasa or Flickr you can display your photos on your TV screen. YouTube is one of the recent additions, becoming available in August.

Devices

You don't have to purchase a NOW TV box. You can watch on Chromecast, many games consoles, Roku, YouView, LG Smart TVs, PCs, mobiles, tablets and more.

NOW TV box

With the same footprint as a beer mat, the Sky NOW TV box turns your television into a Smart TV. For £14.99 you get a small black box, power supply, HDMI cable, and a simple remote control. Setting it up is straightforward. Plug in power, then connect it to your TV (an audio output is available if you need it) and register for an account. The new box offers both Ethernet and WiFi connection, and a processor that is five times faster than the previous model.

If you have a few TVs in the house it's cheap enough to purchase one for each TV, and is a good way to provide Smart functionality for older TV sets.

Chromecast

The new Chromecast device is a simple way to stream content from your phone, tablet, or PC to your TV.

At £30, Chromecast represents one of the most affordable and simplest ways of turning a non-Smart TV with a spare HDMI input from a dumb one into a Smart one. Like the original device, the new Chromecast plugs directly into your TV's HDMI port and streams video from your mobile phone, tablet or PC to your TV.

The only other requirement is a WiFi home network and a mains socket or USB power source (which could be one of the TV's USB sockets). There is no remote control and no user-interface – unlike Google's more sophisticated Smart TV adapter, the Nexus player – since all it does is mirror exactly what is showing on your phone, tablet or PC.

Casting

With casting you're streaming data from your router to your phone or tablet, then back to the router and on to the TV. The process is called 'casting' and works with either the Chrome browser or a number of Chromecast apps, hence the name Chromecast. Compatibility includes Google/Android, PC, iOS and Mac use (albeit with fewer apps).

The new version

Compared with the original Chromecast, the new version offers a range of changes including a new shape. It ditches the dongle-like rigid-stick appearance, in favour of a hockey puck connected by a short flat cable to an HDMI plug, which connects to your TV. This is good because the old model's shape meant it was not guaranteed to fit directly in to some TVs' HDMI sockets. This model has three colour options – black, red and yellow.

Internally, there's a dual-channel 2.4/5GHz WiFi antenna with 802.11ac support, which permits faster streaming and fewer annoying pauses for video buffering. The new version's Chromecast app not only aids the process of setting up and finding other apps such as YouTube and Netflix, it now pre-buffers what you might choose to watch next, based on your previous choices. Universal search is a new, more efficient way of finding content, allowing you to simply enter a title. It's the app equivalent of an EPG that covers all available channels and on-demand providers rather than say just searching BBC or ITV channels.

Set up

This is straightforward on Android devices – just follow the step-by-step instructions on the Chrome website that enable the downloading of software and subsequent process of connecting the device to your WiFi network. Android setup is automated, but iOS is a little more convoluted, requiring a temporary network, generated by the Chromecast.

Operation

There's no on/off button; in order to cast content you just select the HDMI input on your TV and start watching by tapping the dedicated Cast icon that is built-in to any compatible apps – then select the playback device from the list on your network.

Performance

Quality is generally excellent, with buffering proving far less of an issue than the previous version. Dedicated apps are better than watching movies on ITV Player from the Chrome browser, where audio is more prone to lose synch with the picture (with no means of adjusting the lag). Apps also vary in terms of stability but the major ones – Netflix, YouTube, etc. – seem reliable.

A major limitation of Chromecast is that it ties up the use of the device (perhaps your phone or tablet) being used as the source, so taking a phone call or texting whilst watching Amazon Prime is out of the question. But for the relatively low price Chromecast is certainly an appealing proposition.

Google Nexus Player

Following the success of its Chromecast device, we've looked at the company's Nexus Player to see what it has to offer the viewer.

Google takes the premise of its entry-level Chromecast gadget one step further in the guise of the £79 Nexus Player. Rather than, for example, watching movies on Netflix that have been 'cast' from a smartphone, tablet or computer on to a TV screen via the Chromecast dongle, Nexus offers to eliminate the need for a source device out of the equation – but you still have the option to use one if you prefer.

Being a Google product it operates on the Android system, but even devoted Android smartphone and tablet users may wonder if the Android TV operating system is polished enough in its execution.

The hardware
Made by Asus, in the shape of miniature black flying saucer, the Nexus Player is designed to look discreet.

Its specification includes an Intel Atom quad core 1.8GHz processor with 1GB RAM. Wireless networking is provided courtesy of 802.11ac WiFi, which is fast and stable but may be precluded on older routers, and there's no Ethernet option (although a micro USB-to-Ethernet adapter would work). Bluetooth 4.1 is provided for use with the remote control, which has a four-way control and main select button in the centre, plus there are buttons for Voice Search and Back, Play/Pause and Home. An optional game controller is available.

Wired connectivity consists of an HDMI output, a micro USB port and a power socket, all of which are squirrelled away out of sight in a recess so that you only see the cables not the plugs. The lack of an optical audio output may make it slightly more tricky to hook up a soundbar or AV receiver, with HDMI the only audio output option. Another gripe is the meagre storage – 8GB, a mere 5GB of which are available for storing games or apps.

Setup and operation
Setting up is straightforward: you connect to your wireless network, use your Google ID to sign in and wait while it syncs with your account. Oddly, there's no multiple user option. The interface uses Android icons and ranges content across the display in a multi-tiered structure. It's a bright and colourful display with a changing background and responsive display that makes navigation efficient and timely. There's a search bar at the top, with most recently viewed items or recommended shows listed beneath, plus all the major apps and games.

Apps and content
The display shows your installed apps and Google's own Play hubs for games, music and movies. Voice search is done via the microphone built-in to the remote.

Android TV, as Google's portal on your television set, is limited, particularly so for UK users. The Nexus Player isn't a TV tuner, so you don't get free broadcast TV. There's also a fairly sparse offering in terms of apps actually worth having, which is the biggest weakness of the Nexus Player.

The operating system isn't flawless either, and worst of all, the UK's most popular catch-up TV providers aren't catered for so BBC iPlayer, ITV Player, All 4 and Demand 5 are absent (for now, at least). The device's biggest strength at the moment is probably its games repository. If you've bought movies or TV from Google, then Nexus Player gives you access. The same applies to Google Music - both of which are rarer on set-top boxes (but, again, are available via Chromecast at a cut of the price).

The Nexus Player supports Google Cast - so you can 'cast content from phone or tablet to TV, but you can do the same thing using a Chromecast dongle which only costs £30.

BBC iPlayer

With BBC iPlayer you can catch up with the television and radio programmes from the BBC that have been broadcast during the last 30 days. You can view live TV or listen to live radio, and also download programmes to an ever-increasing range of devices.

Programme window: you can stream or download from here and select HD or SD.

Since its launch, in 2007, iPlayer has notched up billions of requests. You can now access the service on a wide range of products, but it is optimised for use on PC, tablet, mobile and TV devices.

BBC iPlayer is continually evolving. This year the BBC has been rolling out new features as part of its drive to 'personalise' the user experience. The on-demand TV service is changing to keep pace with popular streaming services such as Netflix. The service is adding a Live Restart feature – which already works on desktop computer versions of the app – to smart TVs, enabling users to jump back to the beginning of a show at any time during the live broadcast. Cross-device pause and resume – the ability to pause a stream on one device and then pick it up on another – has also been added, as has the My Programmes section, which already works on the iPlayer website, but will now extend to mobile devices. You need to register to make use of some of these new features.

Audio Description (AD) functionality is now available on iPlayer on TV platforms – as well as computers, mobiles and tablets.

Watching live TV

With iPlayer you can stream content to your device and watch BBC One, Two, Three, Four, News Channel, BBC Parliament, CBBC, CBeebies, BBC News and BBC Alba live. While you are watching you can fast forward to any point within a programme and turn subtitles on and off.

Non-BBC content

Limited content from other broadcasters is listed in the TV Guide. Click on the link TV Guide then at the bottom of this page click on the link to Channels from other broadcasters. Clicking on a programme from another broadcaster will take you to its website.

Streaming v downloads

With iPlayer you can select a programme and watch it while it 'streams' to your computer. You can pause playback and then resume watching when you are ready. Depending on the speed of your internet connection you can choose between normal and high quality streaming. However if you are watching on a PC or a mobile device, such as a phone or tablet, and do not have a sufficiently fast connection it is better to download the programme.

Downloads

You can download programmes from BBC iPlayer to watch anywhere, whether you're at home or on the go. So even if you have no WiFi, or your broadband's struggling to download content, you can still watch the programme. Most downloads can be watched any number of times during a 30 day period after broadcast.

To do this on desktop and laptop computers (Macs and PCs) you'll need to install the BBC iPlayer Downloads desktop app from the BBC iPlayer website.

On compatible Android mobiles and tablets, and on iPhones, iPads and iPod Touches, you can download programmes using the BBC iPlayer app. Android users will need to download the BBC Media Player

BBC IPLAYER 43

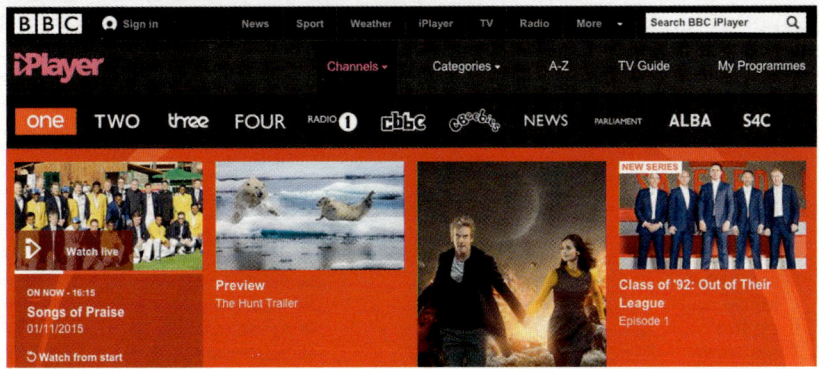

app too. You'll find downloads of the latest software available from Google Play Store, Amazon Appstore and the Apple App Store.

What's available
You'll find live TV and radio as well as hundreds of hours of BBC TV and radio shows from the last 30 days, but it's worth noting that television and radio content has now been separated and are accessed using different versions of the BBC iPlayer app.

Radio listening on iPlayer
BBC iPlayer Radio allows users to listen live to almost all the BBC national and local stations. As with TV, speech and music programmes from across the BBC's national radio stations (such as BBC Radio 4 and BBC Radio 2) and BBC radio stations in Scotland, Wales and Northern Ireland are now available to stream for 30 days on BBC iPlayer Radio. 30 day availability for programmes from regional radio stations and the World Service is currently rolling-out.

High-definition content
Programmes that have been broadcast on BBC One HD and BBC Two HD are usually available to watch in HD (high definition). Sometimes, a programme may not be available in HD due to rights restrictions. Users can select the Watch in HD logo on the programme page. To turn HD off, for example if you have a slow connection, simply toggle the HD on/off switch on the programme page. To download in HD, select Watch in HD before selecting your Download Options.

Can you watch when you're abroad?
Rights agreements mean that BBC iPlayer television programmes are only available to users to stream in the UK, however programmes downloaded in the UK onto mobile devices can be watched offline abroad. Some overseas users resort to IP address hiding services and the use of VPN computer networks to access iPlayer while abroad – and this is something that the BBC started to restrict during 2015.

iPlayer on Smart TVs
iPlayer is now a pre-loaded feature on many Smart TVs and set-top boxes. Alternatively the app can often be downloaded from the manufacturer's app store. If you have an older set you can add iPlayer by choosing a set-top box that supports it.

Computer requirements
You'll need an internet connection that delivers at least 2Mbps of sustained bandwidth for SD content, and at least 3.2Mbps for HD content. If you're having buffering issues while streaming try switching to SD (standard definition). If problems remain contact your internet service provider.

Do you need a TV licence?
You don't need a licence to watch TV on-demand content on iPlayer - for example programmes that you have downloaded, or if you are watching iPlayer recordings from the last 30 days.

However if you are watching a streaming live TV channel, you do. Your household licence will cover streaming content on a mobile device in and out of the home.

Further information
www.bbc.co.uk/iplayer

Television Viewer's Guide

FREEVIEW CHANNELS

Freeview Multiplexes and coverage

Freeview channels are grouped together for transmission in what are called multiplexes (Mux). If you can receive a multiplex you should pick up all its channels. Your location: England, Wales, Scotland, Northern Ireland or the Channel Islands (E, W, W, NI and CI in the table), and whether your aerial is pointed at a main transmitter will also determine which channels you can receive. Relay transmitters do not carry the commercial multiplexes.

	Service	Genre	Mux	E	W	S	NI	CI
1	BBC ONE	Gen Ent	PSB1	✓				✓
1	BBC ONE NI	Gen Ent	PSB1				✓	
1	BBC ONE Scot	Gen Ent	PSB1			✓		
1	BBC ONE Wales	Gen Ent	PSB1		✓			
2	BBC TWO	Gen Ent	PSB1	✓				✓
2	BBC TWO NI	Gen Ent	PSB1				✓	
2	BBC TWO Scot	Gen Ent	PSB1			✓		
2	BBC TWO Wales	Gen Ent	PSB1		✓			
3	ITV	Gen Ent	PSB2	✓				✓
3	ITV Wales	Gen Ent	PSB2		✓			
3	STV	Gen Ent	PSB2			✓		
3	UTV	Gen Ent	PSB2				✓	
4	Channel 4	Gen Ent	PSB2	✓		✓	✓	✓
4	S4/C	Gen Ent	PSB2		✓			
5	Channel 5	Gen Ent	PSB2	✓	✓	✓	✓	✓
6	ITV 2	Gen Ent	PSB2	✓	✓	✓	✓	✓
7	BBC THREE	Gen Ent	PSB1	✓	✓	✓	✓	
8	Local TV (if avail)	News	L_MUX	✓			✓	
8	BBC ALBA	Gen Ent	PSB1			✓		
8	Channel 4	Gen Ent	PSB2		✓			
9	BBC FOUR	Gen Ent	PSB1	✓	✓	✓	✓	
10	ITV3	Gen Ent	COM4	✓	✓	✓		
10	ITV3	Gen Ent	PSB2				✓	
11	Pick	Gen Ent	COM5	✓	✓	✓	✓	
12	Dave	Gen Ent	COM5	✓	✓	✓	✓	
13	Channel 4+1	Gen Ent	PSB2	✓	✓	✓	✓	
14	More Four	Gen Ent	PSB2	✓	✓	✓	✓	
15	Film4	Gen Ent	PSB2	✓	✓	✓	✓	
16	QVC	Gen Ent	COM4	✓	✓	✓	✓	
17	Really	Gen Ent	COM5	✓	✓	✓	✓	
18	4Music	Gen Ent	COM6	✓	✓	✓	✓	
19	Yesterday	Gen Ent	COM6	✓	✓	✓	✓	
20	Drama	Gen Ent	COM4	✓	✓	✓	✓	
21	5 USA	Gen Ent	COM4	✓	✓	✓	✓	
22	Ideal World	Gen Ent	COM6	✓	✓	✓	✓	
23	Local TV (if avail)	News	L_MUX			✓	✓	
24	ITV4	Gen Ent	PSB2	✓	✓	✓	✓	✓
25	Dave ja vu	Gen Ent	COM 6					
26	ITVBe	Gen Ent	COM4					
27	ITV2 +1	Gen Ent	COM4					
28	E4	Gen Ent	COM4	✓				
28	E4	Gen Ent	PSB2		✓	✓	✓	✓
29	E4+1	Gen Ent	COM5	✓	✓	✓	✓	
30	5*	Gen Ent	COM4	✓	✓	✓	✓	
31	Spike	Gen Ent	COM6	✓	✓	✓	✓	
32	Movie Mix	Gen Ent	COM5	✓	✓	✓	✓	
33	ITV +1	Gen Ent	PSB2	✓	✓			
33	STV+1	Gen Ent	PSB2			✓		
33	UTV+1	Gen Ent	PSB2				✓	
34	ITV3+1	Gen Ent	COM6	✓	✓	✓		
35	QVC Beauty	Gen Ent	COM6	✓	✓	✓	✓	
36	Create & Craft	Gen Ent	COM5	✓	✓	✓	✓	
37	QUEST	Gen Ent	COM4	✓	✓	✓	✓	
38	QUEST +1	Gen Ent	COM6	✓	✓	✓	✓	
39	The Store	Gen Ent	COM4	✓	✓	✓	✓	
40	Rocks & Co 1	Gen Ent	COM5	✓	✓	✓	✓	
41	Food Network	Gen Ent	COM5	✓	✓	✓	✓	
42	Travel Channel	Gen Ent	COM6	✓	✓	✓	✓	
43	Gems TV	Gen Ent	COM5	✓	✓	✓	✓	
44	Channel 5+1	Gen Ent	COM4	✓	✓	✓	✓	
45	Film4+1	Gen Ent	PSB3	✓	✓	✓	✓	
46	Challenge	Gen Ent	COM5	✓	✓	✓	✓	
47	4seven	Gen Ent	COM6	✓	✓	✓	✓	
48	movies4men	Gen Ent	COM5	✓	✓	✓	✓	
49	TJC	Gen Ent	COM4	✓	✓	✓	✓	
50	movies4men+1	Gen Ent	COM7	✓				
51	TG4	Gen Ent	RNI_1				✓	
52	RTÉ One	Gen Ent	RNI_1				✓	
53	RTÉ Two	Gen Ent	RNI_1				✓	
54	Heart TV	Gen Ent	G_MAN	✓				
55	Channel 5+24	Gen Ent	COM4	✓	✓	✓	✓	
56	CAPITAL TV	Gen Ent	G_MAN	✓				
58	VIVA	Gen Ent	COM7	✓	✓	✓	✓	
59	BT Showcase	Gen Ent	COM6	✓	✓	✓	✓	
67	Chart Show TV	Gen Ent	G_MAN	✓				
61	True Entertain.	Gen Ent	COM4	✓	✓	✓	✓	
62	ITV4+1	Gen Ent	COM4	✓	✓	✓	✓	
63	Community	Gen Ent	COM7	✓	✓	✓	✓	
64	CBS Action	Gen Ent	COM6	✓	✓	✓	✓	
65	TBN UK	Gen Ent	COM5	✓	✓	✓	✓	
66	CBS Reality	Gen Ent	COM4	✓	✓	✓	✓	
68	TruTV	Gen Ent	COM5	✓	✓	✓	✓	
69	TruTV+1	Gen Ent	COM5	✓	✓	✓	✓	
70	Horror Channel	Gen Ent	COM6	✓	✓	✓	✓	
71	Motors TV	Gen Ent	COM7	✓	✓	✓	✓	
72	ITVBe+1	Gen Ent	COM4	✓	✓	✓	✓	
73	YourTV	Gen Ent	COM5	✓	✓	✓	✓	
74	CBS Drama	Gen Ent	COM4	✓	✓	✓	✓	
76	Jewellery Maker	Gen Ent	COM6	✓	✓	✓	✓	
77	Rishtey Europe	Gen Ent	COM7	✓	✓	✓	✓	
78	YourTV+1	Gen Ent	COM6	✓	✓	✓	✓	
80	Showbiz TV	Gen Ent	G_MAN	✓				
81	Talking Pictures	Gen Ent	COM7	✓	✓	✓	✓	
101	BBC 1 Scot HD	HD	PSB3			✓		
101	BBC 1 Wales HD	HD	PSB3		✓			
101	BBC ONE HD	HD	PSB3	✓				✓
101	BBC ONE NI HD	HD	PSB3				✓	
102	BBC TWO HD	HD	PSB3	✓	✓	✓	✓	✓
103	ITV HD	HD	PSB3	✓	✓			
103	STV HD	HD	PSB3			✓		
103	UTV HD	HD	PSB3				✓	
104	Channel 4 HD	HD	PSB3	✓	✓	✓	✓	

Television Viewer's Guide

FREEVIEW CHANNELS 45

Service	Genre	Mux	E	W	S	NI	CI
105 BBC THREE HD	HD	PSB3	✓	✓	✓	✓	✓
106 BBC FOUR HD	HD	COM7	✓	✓	✓	✓	
107 BBC NEWS HD	HD	COM7	✓	✓	✓	✓	
108 Al Jazeera HD	HD	COM7	✓	✓	✓	✓	
109 Channel 4+1 HD	HD	COM7	✓	✓	✓	✓	
110 4seven HD	HD	COM7	✓	✓	✓	✓	
111 QVC+1 HD	HD	COM8	✓	✓	✓	✓	
112 QVC Beauty HD	HD	COM8	✓	✓	✓	✓	
120 CBBC Channel	Children's	PSB1	✓	✓	✓	✓	✓
121 CBeebies	Children's	PSB1	✓	✓	✓	✓	✓
122 CITV	Children's	COM4	✓	✓	✓	✓	
123 CBBC HD	Children's	PSB3	✓	✓	✓	✓	✓
124 CBeebies HD	Children's	COM7	✓	✓	✓	✓	
125 POP	Children's	L_MUX	✓	✓	✓	✓	
126 Tiny Pop	Children's	COM5	✓	✓	✓	✓	
130 BBC NEWS 24	News	PSB1	✓	✓	✓	✓	✓
131 BBC Parliament	News	PSB1	✓	✓	✓	✓	✓
132 Sky News	News	COM5	✓	✓	✓	✓	
133 Al Jazeera Eng	News	COM6	✓	✓	✓	✓	
134 Al Jazeera Arabic	News	COM7	✓	✓	✓	✓	
135 RT Russia Today	News	COM6	✓	✓	✓	✓	
171 Television X	Adult	COM4	✓	✓	✓	✓	
172 SmileTV2	Adult	COM6	✓	✓	✓	✓	
173 SmileTV3	Adult	COM5	✓	✓	✓	✓	
174 Babestation	Adult	COM6	✓	✓	✓	✓	
175 Party	Adult	COM5	✓	✓	✓	✓	
176 Blue	Adult	COM5	✓	✓	✓	✓	
177 Babestation 2	Adult	COM5	✓	✓	✓	✓	
182 xxXpanded TV	Adult	COM4	✓	✓	✓	✓	
183 AdultXXX	Streamed	COM4	✓	✓	✓	✓	
184 Adult SinTV	Streamed	COM4	✓	✓	✓	✓	
200 BBC Red Button	Text service	PSB1	✓	✓	✓	✓	✓
201 Holidays TV	Text service	COM4	✓	✓	✓	✓	
202 Rabbit	Text service	COM4	✓	✓	✓	✓	
203 Gay Rabbit	Text service	COM4	✓	✓	✓	✓	
204 1-2-1 Dating	Text service	COM4	✓	✓	✓	✓	
207 Kiss Me TV	Text service	COM5	✓	✓	✓	✓	
208 Proud Dating	Text service	COM5	✓	✓	✓	✓	
225 Connect 1	Streamed	COM6	✓	✓	✓	✓	
226 CCTV	Streamed	COM6	✓	✓	✓	✓	
227 Connect 2	Streamed	COM6	✓	✓	✓	✓	
231 Racing TV	Streamed	COM6	✓	✓	✓	✓	
234 Connect 4	Streamed	COM4	✓	✓	✓	✓	
239 Sonlife	Streamed	COM4	✓	✓	✓	✓	
240 Motors	Streamed	COM5	✓	✓	✓	✓	
241 TVPlayer	Streamed	COM4	✓	✓	✓	✓	
242 Vintage TV	Streamed	COM5	✓	✓	✓	✓	
244 Vision TV	Streamed	COM5	✓	✓	✓	✓	
245 Planet Knowledge	Streamed	COM7	✓	✓	✓	✓	
246 JSTV	Streamed	COM4	✓	✓	✓	✓	
247 kykNET	Streamed	COM4	✓	✓	✓	✓	
248 Africa Free TV	Streamed	COM4	✓	✓	✓	✓	
249 Huda TV	Streamed	COM4	✓	✓	✓	✓	
250 Synapse 2	Streamed	COM4	✓	✓	✓	✓	
251 Synapse 3	Streamed	COM4	✓	✓	✓	✓	
252 ABN TV	Streamed	COM4	✓	✓	✓	✓	
253 Passion TV	Streamed	COM4	✓	✓	✓	✓	
254 Showcase TV	Streamed	COM4	✓	✓	✓	✓	
601 BBC Red Button 1	Interactive	PSB1	✓	✓	✓	✓	✓
602 BBC Red Button 2	Interactive	PSB1	✓	✓	✓	✓	✓
603 BBC Red Button 3	Interactive	PSB1	✓	✓	✓	✓	✓
700 BBC Radio 1	Radio	PSB1	✓	✓	✓	✓	✓
701 BBC R1X	Radio	PSB1				✓	✓
702 BBC Radio 2	Radio	PSB1	✓	✓	✓	✓	✓
703 BBC Radio 3	Radio	PSB1	✓	✓	✓	✓	✓
704 BBC Radio 4	Radio	PSB1	✓	✓	✓	✓	✓
705 BBC R5L	Radio	PSB1	✓	✓	✓	✓	✓
706 BBC R5SX	Radio	PSB1					
707 BBC 6 Music	Radio	PSB1					
708 BBC Radio 4 Ex	Radio	PSB1					
709 BBC Asian Net.	Radio	PSB1					
710 BBC World Sv.	Radio	PSB1					
711 The Hits	Radio	COM6					
712 KISS FRESH	Radio	COM5					
713 KISS	Radio	COM5					
714 KISSTORY	Radio	COM5					
715 Magic	Radio	COM5					
716 heat	Radio	COM6					
717 Kerrang!	Radio	COM5					
718 Smooth Radio	Radio	COM6					
719 BBC R Scotland	Radio	PSB1			✓		
719 BBC R Ulster	Radio	PSB1				✓	
719 BBC R Wales	Radio	PSB1		✓			
719 BBC Local Radio (selected areas)		PSB1	✓				
720 BBC R n Gaidheal	Radio	PSB1			✓		
720 BBC R Cymru	Radio	PSB1		✓			
720 BBC R Foyle	Radio	PSB1				✓	
720 BBC Local Radio (selected areas)		PSB1	✓				
721 BBC Local Radio (selected areas)		PSB1	✓				
722 BBC Local Radio (selected areas)		PSB1	✓				
723 talkSPORT	Radio	COM5	✓	✓	✓	✓	
724 Capital FM	Radio	COM4	✓	✓	✓	✓	
725 Premier Radio	Radio	COM6	✓	✓	✓	✓	
726 U105	Radio	PSB2				✓	
727 Absolute Radio	Radio	COM4	✓	✓	✓	✓	
728 Heart	Radio	COM4	✓	✓	✓	✓	
729 RTÉ RnaG	Radio	RNI_1				✓	
730 Insight Radio	Radio	COM5	✓	✓	✓	✓	
731 Classic FM	Radio	COM5	✓	✓	✓	✓	
732 LBC	Radio	COM5	✓	✓	✓	✓	
733 Trans World Radio	Radio	COM5	✓	✓	✓	✓	

Multiplex Names
PSB1 - Public Service - run by the BBC (also known as BBC A)
PSB2 - Public Service - owned by ITV and Channel 4
PSB3 - Public Service - run by the BBC (also known as BBC B)
COM4 - Commercial - owned by ITV (also known as SDN)
COM5 - Commercial - run by Arqiva (also known as Arqiva A)
COM6 - Commercial - run by Arqiva (also known as Arqiva B)
COM7 - HD multiplex - offers coverage to approx 70% of the population.
COM8 - HD multiplex - offers coverage to approx 70% of the population.
G_MAN - Small scale commercial multiplex covering Manchester
RNI_1 - RTE's multiplex covering Northern Ireland
L_MUX - Local multiplex available in selected areas, carrying a dedicated local news and feature service, plus two national syndicated services

Television Viewer's Guide

Freesat channels

Freesat, the UK digital TV satellite service, launched on 6th May 2008. Run by the BBC and ITV, it currently carries over 200 television and radio channels.

The service makes use of the SES Astra satellites (Astra 2E/2F/2G) located at 28.2 °E. These are the same group of satellites used for the Sky pay-TV platform.

High-definition
One of Freesat's key features is access to subscription-free high-definition TV. A good range of HD channels is now available.

New channels
New channels join the service on a regular basis and should appear on the Freesat Electronic Programme Guide (EPG) automatically. Check the Freesat website - www.freesat.co.uk for the latest information.

Adding channels
Freesat's EPG shows channels that have obtained a listing in the EPG. However, additionally, there are over 200 free-to-air TV and radio channels that are not on the EPG but can be manually added to a Freesat set-top box.

ITV regions
Freesat selects the appropriate ITV region based on the postcode entered during set top box installation. Most boxes can tune manually to other ITV regions. Consult the manual for your set-top box for information about how to do this. You'll need to manually enter the frequency information for channels you wish to view. (See page 48.)

Further information
You can find a list of free-to-air channels at: http://en.wikipedia.org/wiki/List_of_free-to-air_channels_in_the_UK

Freesat
www.freesat.co.uk/

Standard-definition channel | High-definition channel

ENTERTAINMENT		
101	BBC One	
102	BBC Two HD	
103	ITV	
104	Channel 4	
104	S4C (Wales)	
105	Channel 5	
106	BBC Three HD	
107	BBC Four HD	
108	BBC One HD	
109	BBC Two	
110	BBC Alba	
112	ITV+1	
113	ITV2	
114	ITV2+1	
115	ITV3	
116	ITV3+1	
117	ITV4	
118	ITVBe	
119	ITV HD	
120	S4C Digidol	
121	Channel 4+1	
122	E4	
123	E4+1	
124	More4	
125	More4+1	
126	Channel 4 HD	
127	4seven	
128	Channel 5+1	
129	5USA	
130	5USA+1	
131	5*	
132	5*+1	
133	5*+24	
134	CBS Drama	
135	CBS Reality	
136	CBS Reality+1	
137	CBS Action	
138	Horror Channel	
139	Horror Channel+1	
140	BET	
141	Spike	
142	True Entertainment	
143	More > Movies	
144	Pick	
145	Challenge	
146	Challenge+1	
147	BBC3	
148	BBC4	
149	Food Network	
150	Travel Channel	
151	Food+1	
152	Travel Channel+1	
153	True Drama	
154	TruTV	
155	ITV4+1	
156	More Movies+1	
157	Show Biz	
158	ITVBe+1	
159	Pick+1	
160	PBS America	
161	Your TV	
NEWS AND SPORT		
200	BBC News HD	
201	BBC Parliament	
202	Sky News	
203	Al Jazeera English	
204	Euronews	
205	France 24	
206	Russia Today (RT) HD	
207	CNN	
208	Bloomberg TV	
209	NHK World HD	
210	CNBC	
211	CCTV News	
212	BBC News	
213	Channels24	
MOVIES		
300	Film4	
301	Film4+1	
302	True Christmas	

FREESAT CHANNELS 47

303	True Movies 2		**RADIO CHANNELS**		808	TJC Choice
304	Movies4Men		700	BBC Radio 1	809	TJC
305	Movies4Men+1		701	BBC Radio 1Xtra	810	Ideal Extra
	LIFESTYLE		702	BBC Radio 2	811	Craft Extra
400	Irish TV		703	BBC Radio 3	812	Ideal World
401	Information TV		704	BBC Radio 4 FM	813	Create and Craft
402	Showcase		705	BBC Radio 5 Live	814	Rocks&Co
410	FilmOn TV		706	BBC Radio 5 Sports Ex.	815	Thane Direct
411	Fashion One		707	BBC 6 Music	816	High Street TV
	MUSIC		708	BBC Radio 4 Extra	817	Hochanda
500	Chart Show TV		709	BBC Asian Network		**ADULT**
501	The Vault		710	BBC Radio 4 LW	870	Babestation
502	Chart Show Dance		711	BBC World Service		**REGIONAL**
503	B4U Music		712	BBC Radio Scotland	950	BBC One London
504	STARZ		713	BBC Radio nan Gaidheal	951	BBC One Ch. Islands
505	Vintage TV		714	BBC Radio Wales	952	BBC One East Midlands
506	Heart TV		715	BBC Radio Cymru	953	BBC One East (E)
507	Capital TV		716	BBC Radio Ulster	954	BBC One East (W)
508	Kiss		717	Radio Foyle	955	BBC One North West
509	The Box		718	BBC London 94.9	956	BBC One NE & Cumbria
510	Magic			(London postcodes)	957	BBC One N Ireland
511	Viva		719	Capital	958	BBC One Oxford
512	Chilled		720	Capital xtra	959	BBC One South East
513	NOW Music		721	Classic FM	960	BBC One Scotland
514	Clubland TV		722	Gold	961	BBC One South
	CHILDREN'S		723	Radio X	962	BBC One South West
600	CBBC HD		724	Absolute Radio	963	BBC One West Midlands
601	CBeebies HD		726	Absolute 80s	964	BBC One Wales
602	CITV		729	Jazz FM	965	BBC One West
603	POP		730	Planet Rock	966	BBC One Yorks
604	Kix		731	talkSPORT	967	BBC One Yorks & Lincs
605	Tiny POP		732	Smooth Radio	968	BBC Two England
606	Kix+1		733	Heart	969	BBC Two N. Ireland
607	CBBC		734	LBC	970	BBC Two Scotland
608	CBeebies		750	RTE Radio 1	971	BBC Two Wales
609	Tiny Pop+1		751	RTE 2 FM	972	BBC One HD
	SPECIAL INTEREST		752	RTE Lyric FM	973	BBC One HD Scotland
651	Community Channel		753	Raidió na Gaeltachta	974	Channel 4 London
652	Forces TV		786	BFBS Radio	975	Channel 4+1 London
660	Sony SAB TV		790	Trans World Radio	976	BBC One HD Wales
661	Rishtey			**SHOPPING**	977	ITV1 London
662	Colors		800	QVC	978	BBC One HD N. Ireland
663	Zing		801	QVC Beauty		**ON DEMAND**
690	Inspiration TV		802	QVC Extra	900	On Demand
691	Daystar TV		803	QVC Style	901	BBC iPlayer
692	Revelation		804	GR Shop	903	ITV Player
693	Islam Channel		805	Gems TV	981	BBC Red Button
694	God Channel		806	TV Shop		
695	Sonlife (SBN)		807	Jewellery Maker		

Television Viewer's Guide

Sky channels

Adding channels
Each BBC region now has its own EPG listing - so if you want to pick up BBC local news from another region you can just select it in the EPG. There have been rumours that ITV1 will do the same, but currently if you want to watch an ITV region (other than your local one) you will have to find it manually and then add it to the 'Other Channels' menu.

To do this go to Services, System Setup, Add Channels.

Input the following data:
Symbol Rate is 22, and FEC 5/6 for the frequencies listed below. The H and V will need to be entered - this refers to polarization.
10.758V - London, Granada, Anglia East and Central West
10.832H - Wales, West, Westcountry
10.891H - Border, Meridian S, Meridian SE, Yorkshire W, and Tyne Tees
10.906V - Channel Islands, Scottish TV East, North and West) and Ulster TV.
10.994H - Anglia West, Central South West, Meridian North, Yorkshire East.
11.053H - Central East, Central South
Once you have input the appropriate details, select Find Channels. Store the ones you want as per the onscreen instructions. To view them you will need to select Services - Other Channels.

Owners of Freesat satellite boxes can also tune manually to these ITV regional channels. You will need to put your receiver into non-Freesat mode. This is a function on Freesat satellite receivers allowing viewers to access other free-to-air channels broadcast from the satellites your dish is pointing at, but which are not in the Freesat EPG channel list.

Yellow channels are free-to-air. You don't need a viewing card to watch them.
Pink channels are encrypted and require a viewing card and subscription to the channel.
Grey channels are HD channels. You will need an HD set-top box. Some are free and some are only available by subscription.

Free-to-air | Subscription needed | HD channel

#	Channel	#	Channel	#	Channel
101	BBC One	123	ITV Encore	146	CBS Reality
102	BBC Two	124	FOX	147	CBS Reality +1
103	ITV	125	TLC	148	CBS Action
104	Channel 4	126	MTV	149	CBS Drama
105	Channel 5	127	Comedy Central +1	150	Universal+1
106	Sky1	128	Comedy Central Xtra	151	E!
107	Sky Living	129	Sky 2	152	Pick TV
108	Sky Atlantic	130	Sky 1+1	153	Pick TV+1
109	Watch	131	ITV1+1	154	TLC+1
110	GOLD	132	alibi	155	Really
111	Dave	133	Good Food	156	Lifetime
112	Comedy Central	134	S4C	157	Sony TV
113	Universal	135	Channel 4 +1	158	Drama
114	Syfy	136	E4	159	Comedy Xtra+1
115	BBC Three	137	E4+1	160	Spike
116	BBC Four	138	More4	164	Challenge+1
117	London Live	139	More4+1	165	Fox+
118	ITV2	140	4seven	166	GOLD +1
119	ITV3	141	BBC One HD	167	DMAX
120	ITV4	142	BBC Two HD	170	Sky Atlantic +1
121	Sky Arts	143	BBC ALBA	171	Channel 5 HD
122	Sky Living+1	144	QUEST	172	Real Lives
		145	Challenge	173	Real Lives +1

SKY CHANNELS 49

174	5 USA	239	E! HD	334	Sky Action HD
175	Channel 5+24	240	Home&Health	335	Sky Comedy HD
176	5*	241	Home&Health+	336	Sky Thriller HD
177	Channel 5+1	242	Discovery Shed	337	Sky Drama Rom HD
178	ITV HD	243	Good Food +1	339	Sky ScFi/Horror HD
179	ITVBe	244	Home +1	340	Sky Select HD
180	ITV2+1	245	Watch HD	341	TCM HD
181	Sony TV+1	246	Dave HD	342	Film 4 HD
182	BEN	247	aliba HD	343	Talking Pictures
183	True Drama	248	Food Network		**MUSIC**
184	True entertainment	249	Travel Channel	350	MTV MUSIC
185	more>movies	250	Food Network+1	351	MTV BASE
186	more>movies+1	251	Travel Channel+1	352	MTV HITS
187	BET:BlackEntTv	252	?TV	353	MTV DANCE
188	FOX HD	253	Horse & Country	354	MTV ROCKS
189	propeller	255	Good Food HD	355	MTV CLASSIC
190	5*+1	256	holiday+cruise	356	VH1
191	Irish TV	257	Fashion One	357	VIVA
192	AMC from BT	261	Showcase	358	MTV Live
193	ITV3+1	263	ITV3 HD	359	The Box
194	Syfy+1	264	Forces TV	360	4Music
195	QUEST+1	265	Info TV+1	361	KISS
196	Home	266	ShowBiz TV	362	Smash Hits!
197	Lifetime+1	268	ITV4 HD	363	Magic
198	Tru TV	269	ITVEncore HD	364	Heat
199	OH TV	270	MTV +1	365	Chart Show TV
200	TLC+2	271	tru TV +1	366	The Vault
201	SyFy HD	272	Property +2	367	Scuzz
202	E4 HD	275	Your TV	368	Kerrang!
203	My Channel		**MOVIES**	369	Vintage TV
204	Universal HD	301	Sky Premier	370	Chilled TV
205	Comedy Central HD	302	Sky Premier +1	371	Starz TV
206	ITV4+1	303	Sky Showcase	372	Chartshow Dance
207	ITVBe +1	304	Sky Superhero	374	Flava
208	ITVEncore +1	305	Sky Disney	376	Bliss
210	BBC Three HD	306	Sky Family	378	NOW Music
211	BBC Four HD	307	Sky Action	383	Clubland TV
212	Information TV	308	Sky Comedy	385	Channel AKA
213	Watch +1	309	Sky Thriller	386	MTV Live HD
214	Dave ja vu	310	Sky Drama Rom	387	Heart TV
217	Sky 1HD	311	Sky ScFI/Horror	388	CAPITAL TV
218	Vox Africa	312	Sky Select		**SPORTS**
219	5 USA+1	314	Sky Premier HD	401	Sky Sports News HQ
220	MTV HD	315	Film4	402	Sky Sports 1
221	Sky Living HD	316	Film4 +1	403	Sky Sports 2
222	Sky Atlantic HD	317	TCM	404	Sky Sports 3
223	Sky Arts HD	318	TCM +1	405	Sky Sports 4
225	ITV4 HD	319	horror channel	406	Sky Sports 5
226	alibi +1	320	horror channel+1	407	Sky Sports F1
227	Channel 4 HD	321	True Movies 1	408	Sky Sports News HQ HD
229	E!+1	322	True Movies 2	409	Sky Sports 1 HD
974	Channel 4	323	Sony Movies	410	Eurosport
975	Channel 4 +1	324	Sony Movies+1	411	Eurosport 2
	LIFESTYLE AND CULTURE	325	movies4men	412	Eurosport HD
231	More4 HD	326	mov4men+1	413	BT Sport 1
232	AIT Int'l	327	Movies24	414	BT Sport Europe
233	DMAX+1	328	Movies24+	415	At The Races
234	TLC HD	329	Nollywood	417	BT Sport 2
235	ABN TV	330	Sky Showcase HD	418	MUTV
236	Lifetime HD	331	Sky Superhero HD	419	Extreme Sports
237	Film On TV	332	Sky Disney HD	421	Chelsea TV
238	Property Show	333	Sky Family HD	422	Chelsea TV HD

Television Viewer's Guide

SKY CHANNELS

#	Channel
426	BT Sport/ESPN
427	BT Sport 2 HD
428	Premier Sports
429	Liverpool FC TV
432	Racing UK
433	BT Sport Europe HD
435	Sky Sports 2 HD
436	Sky Sports 3 HD
437	Box Nation
439	Sky Sports 4 HD
440	Sky Sports 5 HD
443	Eurosport 2 HD
447	MOTORS TV UK
451	Sky Sports F1 HD
453	MUTV HD
455	LFCTV HD
457	BT Sport 2 HD
458	BT Sport/ESPN HD
462	Premier HD
490	Box Nation HD
491	Sky Sports Box Office
492	Sky Sports Box Office HD

NEWS

#	Channel
501	Sky News
502	Bloomberg
503	BBC News
504	BBC Parliament
505	CNBC
506	CNN
507	NHK World HD
508	Euronews
509	FOX News
510	CCTV News
511	NDTV 24x7
512	Russia Today
513	FRANCE Eng
514	Al Jazeera Eng
515	CNC World
516	Sky News HD
517	TVC News
518	Russia Today HD
519	ARISE News
570	BBC News HD
571	BON TV
572	TVC News +1
575	Channels24

DOCUMENTARIES

#	Channel
520	Discovery
521	Discovery +1
522	ID
523	Animal Planet
524	Discovery Turbo
525	Discovery Science
526	Nat Geographic
527	Nat Geographic+1
528	Nat Geographic Wild
529	History
530	History +1
531	H2
532	Eden
533	Eden+1
534	PBS America
535	Discovery History
536	Discovery History+1
537	YESTERDAY
538	YESTERDAY+1
539	CommunityChnl
543	Nat Geographic HD
544	Nat Geographic Wild HD
545	History HD
549	Animal Planet +1
551	ID+1
553	CI
554	CI+1
555	CI HD
557	Discovery Science+1
561	Discovery HD
562	Animal Planet HD
563	Eden HD

RELIGION

#	Channel
580	GOD Channel
581	revelation
582	TBN UK
583	DAYSTAR
585	Inspiration
586	LOVEWORLD TV
587	Gospel Channel
588	Word Network
589	EWTN Catholic
590	Faith World TV
591	KICC TV
592	Believe TV
593	Olive TV
594	SonLife
595	Flow TV

CHILDREN'S

#	Channel
601	Cartoon Network
602	Cartoon Network+1
603	Boomerang
604	Nickelodeon
605	Nickelodeon+1
606	Nicktoons
607	DisneyXD
608	DisneyXD+1
609	Disney Channel
610	Disney Channel+1
611	Disney Junior
612	Disney Junior+1
613	CBBC
614	CBeebies
615	Nick Jr.
616	POP
617	Tiny Pop
618	Boomerang + 1
619	Cartoonito
620	Nick Jr. Too
621	CITV
622	Disney XD HD
623	Baby TV
624	CBeebies HD
625	Tiny Pop +1
626	POP +1
627	Kix
628	Disney Junior HD
629	Kix +1
630	Nick Jr+1
631	Disney channel HD
632	NickelodeonHD
633	CBBC HD
634	Cartoon NetworkHD
636	Bommerang HD

SHOPPING

#	Channel
650	QVC
651	JML Direct
652	TJC
653	GR SHOP
654	Ideal World
655	Gems TV
656	Tristar
657	Retail TV
658	High Street TV
659	Best Direct
660	TJC Choice
661	Ideal Extra
662	High Street TV2
663	Hochanda
665	Jewelry Maker
666	High Street TV3
667	TV Warehouse
668	QVC Beauty
669	PaversShoes.tv
670	Thane Direct
671	V Channel
672	Rocks & Co 1
673	QVC Extra
674	Create & Craft
675	Craft Extra
676	The Dept Store
677	The Mall
678	QVC Style

SKY BOX OFFICE

#	Channel
700-719	Sky Box Office
743-744	Sky Box Office
752	SBO HD

INTERNATIONAL

#	Channel
780	B4U Movies
781	B4U Music
782	Sony TV Asia
783	Star Life OK
784	Star Plus
785	PCNE Chinese
786	Bangla TV
787	mta-muslim tv
788	Zee TV
789	Zing
790	Zee Cinema
791	ARY Digital
792	PTV Prime
793	MATV National
794	Abu Dhabi TV
795	& TV
796	TV5 Monde
797	UMP Movies
798	Sony MAX
799	Plus TV
800	HidayatTV

Television Viewer's Guide

SKY CHANNELS 51

801	Record TV		999	Sky	0103	BBC R3
802	ARY News			**ADULT**	0104	BBC R4 FM
803	PTV Global		900	Playboy TV	0105	BBC R5 Live
804	ARY QTV		901	Adult Channel	0106	Classic FM
805	Venus TV		902	Playboy TV Chat	0107	Absolute
806	Islam Channel		903	TelevisionX	0108	talkSPORT
807	GEO UK		904	redhotamateur	0109	Capital FM
808	Star Jalsha		905	redhotmums	0110	Planet Rock
809	STAR Gold		906	Babestation	0111	Heart
810	Sony SAB		907	redhot18s	0112	LBC
811	ABP News		908	Get Lucky TV	0113	Radio X
812	NoorTV		909	Lucky Star	0114	Capital XTRA
813	Peace TV		910	XXXBrits	0115	BBC World Service
814	CHSTV		911	TVX Brits	0116	BBC R Scot.
815	PTC Punjabi		912	Studio 66	0117	BBC R Wales
816	GEO TEZ		913	ExGirlfriends	0118	BBC R Ulster
817	GEO News		914	XpandedTV	0119	BBC Asian
818	AAJTAK		920	xxxbig&bouncy	0120	BBC 6 Music
819	IQRA TV		921	xxxGirlGirl	0121	Gold
825	IQRA Bangla		922	xxxAmateurs	0122	WRN Europe
820	Ummah CHNL		923	XXXGroups	0124	LBC News
821	COLORS		924	playboy TV Hot	0125	Apni Awaz
822	Rishtey		925	XXX 18 Dirty	0128	Smooth
823	Glory TV		926	XXX Mums	0129	Solar Radio
824	Brit Asia TV		930	Climax	0130	Panjab Radio
826	Ahlebait TV		939	Storm	0131	BBC R4 Ex
827	ATN Bangla UK		940	Play TV	0137	BBC R1X
828	Madani Chnl		941	Studio 66	0138	TWR
829	Sikh Channel		942	Sin TV	0139	BBC R n Gael
830	Peace TV Urdu		945	Babes from TV	0143	BBC R4 LW
831	Ahlulbayt TV			**REGIONAL TV**	0144	BBC R5SX
832	Samaa		950	Sky Intro	0146	Smooth Extra
833	Channel i		951	BBC1 Scotland	0147	EWTN
834	TakbeerTV		952	BBC1 Wales	0150	Sukh Sagar
835	Zee Punjabi		953	BBC1 NI	0151	Khushkhabri
836	Sangat		954	BBC1 London	0154	BBC R Cymru
837	Aastha		955	BBC1 NE&C	0160	RTE Radio 1
838	NTV		956	BBC1 Yorks	0164	RTE 2FM
839	Star Plus HD		957	BBC1 Yrks & Lin	0165	RTE Lyric FM
840	Sky News Arabia		958	BBC1 N West	0166	RTE R na G
842	TV 99		959	BBC1 W Mids	0169	Desi Radio
843	Akaal Channel		960	BBC1 E Mids	0178	Kiss
844	HUM Europe		961	BBC1 East (E)	0180	Magic
845	British Muslim		962	BBC1 East (W)	0185	Ahomka RD
847	Safeer TV		963	BBC1 S East	0186	Liberty
848	News 18 India		964	BBC1 South	0188	Absolute 70's
849	Islam TV Urdu		965	BBC1 Oxford	0200	Absolute 80's
850	Dunya News		966	BBC1 West	0201	Absolute 90's
851	Islam Ch Urdu		968	BBC1 CI	0202	Jazz FM
853	& TV HD		970	BBC2 Scotland	0203	Classic Rock
854	Ummah Ch+1		971	BBC2 Wales	0205	Kanshi Radio
855	Al Arabiya		972	BBC2 NI	0207	Rainbow
	GAMING AND DATING		973	ITV1 London	0210	Newstalk
862	SuperCasino		974	Channel 4	0211	BFBS Radio
864	Alexcasino		975	Channel 4 +1	0214	UCB Ireland
872	Gay Network		977	BBC One Scotland HD		
873	Chat Box		978	BBC One Wales HD		
886	Psychic Today		979	BBC One NI HD		
	SPECIALIST		980	BBC Red Button		
899	Sky Intro			**RADIO**		
996	Channel Line-up		0101	BBC R1		
998	Sky Intro		0102	BBC R2		

Television Viewer's Guide

Buying a new television

There are hundreds of TVs available, so how do you go about getting the best one for you? Whether you are shopping in store or online it's easy to get confused. Here is some information to get you started.

Modern TVs can be very thin and are easy to wall mount.

What size screen do you need?

When selecting the screen size which is best for you, it's important to consider the space in your room and how far away you will sit from your TV. There are no hard and fast rules and many manufacturers give a recommended range of distances for their sets. Much depends on your preferences and whether you want an 'immersive' viewing experience by sitting close to the TV. Also consider how the television will fit within the room layout.

A general rule of thumb for HD television sets is that the optimum distance is about two and a half times the screen's diagonal size. For Full HD 1080p and 4K UHD models, which offer better resolution, you may want to sit closer so that you can appreciate the additional detail. Most people choose to sit further away than this, but you may not discern all of the available picture detail.

Viewing angle

You do not need a very wide viewing angle if you are watching the television alone or if your sofa is directly in front of the television. However, if there are several viewers it might affect the experience of the people sitting to one side.

OLED televisions generally offer a better viewing angle than LED because each pixel produces its own light. With LED light is passed through the screen from a single source. Manufacturers tend to be overly generous in their specifications! As a general guide OLED screens offer a viewing angle of around 170 degrees, LCD and LED around 150 degrees.

Screen shape

All new screens have a widescreen viewing ratio (or aspect ratio) of 16:9. This enables you to see the whole picture as programmes are now transmitted in widescreen. A range of curved screens is now available: see page 58.

Wall mounting

Whether you wall mount the television, use it on a stand, or on a piece of furniture depends on personal preference and the size of the screen chosen. With lighter screens a wide range of options are available with fixed, tilting and fully articulating wall mounts.

LED or OLED?

Now that plasma TV screens have gone from the market the choice of viewing screen lies between the several types of LED TVs (which use LCD, Liquid Crystal display, to split LED light into colour dots) and OLED types. Apart from price OLED is superior to LED in every way, and should be chosen if it can be afforded; small screens are only available in LED type. OLED panels can be made thinner than LED ones if that is important to you, and both types can be had in curved form – a questionable advantage.

You will find more information about the different screen types on page 56.

HIGH DEFINITION OPTIONS

The introduction of high-definition (HD) has been one of the greatest changes in TV broadcasting in recent years. HD images are crisper and offer more detail.

BUYING A NEW TV   53

High-definition broadcasts come from Freesat, Sky, Freeview, Virgin cable, and via BBC iPlayer and other catch-up services. They are also available from games consoles: the Sony PlayStation 3 and 4, Microsoft Xbox 360 and Nintendo Wii.

Blu-ray players provide a source of high-definition pictures and digital surround sound. The internet also carries HD content, but you'll ideally need 4Mbps download speed or better to watch it. You can even generate your own HD content with an HD camcorder.

HD television comes in four different formats with the quality steadily improving.

720p

720p uses a 'progressive' system which displays each frame of the image as a whole. 720p is generally accepted as being better at handling fast motion and is thus best suited for news, sports and similar programmes, while 1080i performs better for very fine detail: ideal for documentary, art and nature programmes.

1080i

1080i (currently used by Sky) uses an 'interlaced' system which breaks the image into two fields and displays 'odd' and 'even' fields alternately.

In comparison tests, 1080i appeared to offer slight advantages over 720p in terms of sharpness, smoothness (less jagged curves and angles) and colour toning.

Full HD 1080p

Large LCD and OLED screens can display 1080p images - often called Full HD. Although not yet broadcast because of its large bandwidth demand, 1080p content is available from Blu-ray players and offers marginally better quality images than 1080i.

Ultra HD/4K

So named because the screens offer a resolution of around 4000 pixels across the screen, UHD has overtaken 1080p in the high definition stakes. An Ultra HD image has about four times the resolution of Full HD. However it is unlikely that you will see any noticeable improvements unless you are planning to buy a screen larger than 55-60 inches.

Even then, we would advise you to wait a while before you rush out to buy a UHD TV model. You'll pay quite a lot more; but the real issue is that there is little 4K content available to watch. You'll find no air-based TV channels offering it; in the meantime you will have to rely on upscaled video or the internet where it's available (to those with 20Mbps or faster links) from Amazon and Netflix.

The new HDMI connection version 2.0 is required to handle the large amounts of data that needs to be transferred with 4K.

What do you need to watch HD?

At present every TV set on the market larger

Television Viewer's Guide

BUYING A NEW TV

than 20 inches is HD-capable, and there are scores of HD broadcast channels available on Freeview, Freesat, Sky and Virgin. You'll find further information about high-definition TV on page 62.

3D TELEVISION

You will find 3D built-in to most top-end TVs. Moreover the 2D capability of 3D TVs is not affected in any negative way by 3D capability. Look at 3D as a feature that you may or may not use.

Despite the best efforts of the TV industry, 3D has failed to take off. Viewers are not keen to wear the glasses necessary to watch current 3D programmes, and the range of quality 3D content is limited. Subscription-based 3D content is available on Sky, and there are several hundred Blu-ray 3D titles available, but broadcast content is very limited and is unlikely to improve any time soon. In fact the BBC and sports channel ESPN have suspended their 3D activities for the present.

Active vs Passive 3D

There are two types of 3D TV on the market today, active and passive. Both require glasses, and both will work with broadcast 3D content and Blu-ray players.

Active

In most cases active screens offer higher definition than passive types. We generally recommend them, particularly for the better quality Blu-ray 3D pictures they can produce. The biggest disadvantage of active systems is the cost of the battery-powered glasses, about £100 a pair, expensive if you have a large family or many friends who want to watch. Glasses are not always provided with active 3D TV sets.

Passive

Much like the cinema experience, passive 3D uses polarising, non-powered glasses, which are cheap to buy and replace. This is great if you have a family, as the glasses can get broken by over-enthusiastic youngsters! See page 61 for information about 3D.

WHICH PLATFORM TO CHOOSE

All TVs on sale in the UK now have Freeview tuners, generally HD built-in. There's no need to purchase a separate set-top box to pick up digital television. A few sets incorporate Freesat HD tuners.

Whether to choose Freeview or Freesat, which are free, or Sky or Virgin, the main pay-TV providers, depends predominantly on the content you would like to watch. If you are a keen sports or movie fan, or there is a particular channel you like that is not available on Freeview or Freesat, then Sky or Virgin, or one of the hybrid services such as BT TV or TalkTalk may be worth considering - but look at the free options first.

However it can be confusing when trying to decide which TV platform to choose, as the pay-TV providers offer packages of services including TV along with phone and broadband subscriptions. It is worthwhile checking the comparison sites to see where savings can be made if you are prepared to obtain your TV, phone and broadband from one provider.

You'll find detailed information about the main TV platforms such as Freeview, Freesat, Sky and Virgin elsewhere in this guide.

SOUND QUALITY

As TVs have become progressively thinner, set makers have found it difficult to position speakers at the bottom and sides of the screen. They're now generally banished to the rear of the cabinet, pointing into the wall behind the TV set, or the curtains. The result is very poor tinny sound, with little bass reproduction.

Most TVs feature simple stereo sound or a system that produces a 'virtual' surround effect. This is fine for general watching, but if you wish to improve the sound you have several options.

Soundbars

Soundbars comprise two or more (left/right stereo) loudspeakers in a bar-shaped TV-external box which can form a plinth for the TV to stand on; sit on the stand in front of the screen; or be wall-mounted behind it. They contain their own mains-powered amplifiers and are simple to set up, needing only a single cable to connect to the TV. They often have a separate subwoofer delivering the bass.

With adequately large speakers, built-in amplifiers and solid cabinets, soundbars offer a much higher quality of sound than TV sets. Prices range from under £100 to over £2000.

BUYING A NEW TV — 55

All-in-one home cinema systems
Most commonly available in 5.1 surround sound set-ups, these are often sold in systems that include a Blu-ray player. Prices range from £220 to over £1500.

The 5.1 relates to the audio channels and loudspeaker arrangement - five channels for normal-range speakers (right, centre, left, back right surround, back left surround), and one channel for the subwoofer (low-frequency effects).

Increasingly WiFi is used to connect to the speakers, avoiding the need to run wires all round the room. You'll find further information about TV sound on page 66.

SMART AND INTERNET TV

Smart TV is also referred to as Internet TV or Connected TV. Many of the latest TVs are capable of connecting to the internet. It is a feature, rather like 3D capability, that you do not have to use.

If you want the Smart features of a new TV on an older model you can simply connect your TV to a set-top box with Smart functionality. Many HD Freeview and Freesat boxes, and Blu-ray players provide a range of Smart functions, as does the latest equipment from Sky and Virgin.

There is an increasing range of devices that will enable you to do this too. See the information elsewhere in this guide.

What can Smart TVs do?
Essentially they enable you to access the internet, but you'll need a broadband connection to do so. Each provider has its own version of this technology so you will find different models vary in what they offer.

The best-used feature of a Smart TV is its provision of catch-up TV services such as BBC iPlayer. A whole range of premium programming and films are available.

Social networking is something found on some TVs and Smart set-top boxes. Full web browsing, as you would on a personal computer, is available on a few models but does not enable you to browse as well as you would with a personal computer. For one thing TVs lack the power of a PC, may be more sluggish, and a remote control is not as effective in use as a keyboard.

To enable the Smart TV functions the TV has to be connected to the internet. You can do this by hooking it up to your router and broadband, either via an Ethernet cable, or wirelessly over a home network via WiFi.

WHERE TO BUY TV PRODUCTS

Retailers compete for your business in several different guises, each with pros and cons.

Internet
Purchasing online has rapidly increased in the UK, with about 48 percent of home electrical equipment now purchased in this way. Its advantages are low prices and a wide choice of equipment. Delivery can be quick, even on the next day for an additional charge.

Detailed specifications and sometimes advice on choice and suitability is provided on websites too. John Lewis has competitive prices and at present a free extended warranty on electrical and electronic goods. The main disadvantage of web buying is that there's no opportunity to 'touch and feel' the equipment.

Independent dealers
The traditional TV retailer usually has the highest level of accessible product knowledge; offers the opportunity to examine and appraise the equipment; and may even be able to provide same-day delivery of items in stock. On the negative side prices tend to be higher than those of other vendors, and the product range more limited.

Multiple stores
National chain stores such as Currys/PC World may not be able to offer the product knowledge of many independent dealers but generally have in stock, and on display, a wider product range. Prices tend to fall between those of internet and independent traders. They combine internet and showroom commerce.

Supermarkets
Large branches of supermarkets stock TV and audio goods to take away at competitive prices, but may not have the product knowledge available elsewhere. They also trade on the internet. It is worth checking as you can sometimes find more competitive prices and better offers via their websites.

Television Viewer's Guide

Television types

Prices of LED flat screen televisions have come down a lot in the last few years. LED TV uses variable light polarising LCD (Liquid Crystal Display) technology; the differences lie in the backlighting method. They have taken over from earlier LCD TVs, and you will usually find these screens described as LED.

Plasma	LED	Quantum Dot
Plasma TVs are no longer on sale in the UK, and are only included here for comparison purposes. In their time they were held to give the most 'cinematic' pictures amongst direct-view screens, and reached an advanced level of technology, especially in top-end Panasonic and Pioneer models. **How does a plasma screen work?** Tiny cells held between two panels of glass hold a gaseous mixture. The gas in the cells is electrically turned into plasma which excites phosphors to emit light. **Advantages** Plasmas often give deeper blacks, and have richer colours than LCD screens; they have a wider viewing angle, less motion blur and short image lag. **Disadvantages** Plasma screen images are not as bright as those of an LCD TV and work best in rooms with lower levels of lighting. Glare can be a problem because of the reflective glass screens used. Plasma TVs are more fragile and difficult to handle than LCD sets, and can be affected by screenburn. Power consumption is relatively high.	**How does an LED screen work?** LED TVs use millions of colour LCD shutters to filter the light from a very bright and even light source, produced by a matrix of LEDs (light emitting diodes) at the back of the screen. **Advantages** Unaffected by screenburn, LED panels can provide a brighter image than plasma, good for viewing in bright ambient light. Power consumption is lower than that of similarly-sized plasma screens. **Disadvantages** Backlight leakage can prevent the display of a really dark black level; and uneven illumination and 'cloudiness' may show up on cheaper models. Image lag may be present, to cause comet-tailing on fast-moving images and make play difficult in fast-action TV games. Check these out in the shop if possible - much depends on the price. **Prices** LED TVs have a very wide price span in the market, ranging from budget supermarket specials to excellent home-cinema standard models by top manufacturers. 40 inch models can be had for under £250, while 60-65 inch LED screens range up to £5000.	**Both Samsung and LG** have begun to introduce quantum dot technology to their screens. Quantum dots are tiny semiconducting nanocrystals, ranging in size from 2-10nm (nanometres). They are able to fluoresce (emit light) when irradiated by light of any colour. Their colour depends on their size and is very pure in spectral terms. A quantum dot sheet is interposed between the (now just blue) backlight and the LCD layer in an LED screen assembly to 'purify' the light and make the colour filter dots more efficient in their operation. **Advantages** Quantum Dot screens are brighter than those of ordinary LED types, offering a wider range of colours – especially important in HDR TVs – and a more vivid, vibrant image. Their power consumption is also reduced. **Disadvantages** They are more expensive than conventional LED models. Their ability to provide good black levels is not as good as OLED screens. Even so they are worth consideration if picture quality is important to you and you cannot afford OLED prices.

TELEVISION TYPES   57

The vast majority of TVs for sale now are LED types, which come in many variants, some of which are described below. In this competitive field price is a very good indicator of picture quality in any given screen size.
OLED screens provide the best pictures, but at a high price.
Smart technology is available with both screen types, and is a feature that you may or may not want to use.

OLED	3D	Projectors
OLED is a technology that has entered the mainstream television screen market relatively recently. Standing for Organic Light Emitting Diode, OLED displays offer the best performance of all types of TV screen. **Who makes OLED sets?** Sony pioneered OLED TV six years ago, and continues to develop LED technologies. At present OLED TVs are available in 55-78 inch size in the UK from makers LG and Panasonic. **Advantages** OLEDs need no backlight to function. So they can display deep black levels, draw far less power, and can be much thinner (only 3mm thick) and lighter than an LCD panel. OLED displays also achieve a much higher contrast ratio than LCD monitors. A quicker response time than standard LCD sets imparts smooth and natural motion. **Disadvantages** Mainly the cost, but the price of OLED screens has dropped steeply in the last year. A 55 inch model can be purchased for £1700 upwards, while 65 inch can be found for less than £5000. They all offer 3D features, and most have curved screens.	**3D television** could be a factor in your choice if you are contemplating buying a new TV. The technology has not really caught on, but it is included as standard with many screens. **Can existing TVs show 3D?** Ordinary 2D screens cannot be adapted to 3D, but you can watch 2D television on a 3D screen. Presented with a 3D feed, e.g. from Sky, an ordinary set will blank out. **System choice** 3D screens use either 'Active Shutter' systems or 'Passive' technology. Each has its pros and cons. See page 61 for further information. Don't forget to factor in the cost of glasses (£80-£120 each) if you choose to purchase an 'active' system. **Who makes 3D sets?** Every major manufacturer has 3D screens in its product range, and the premium for 3D (over 2D types) is quite small. Active 3D sets are a little more expensive than passive types. **Prices** 3D display has become standard in mid and high-range models from all manufacturers. See the LED column here, and check in store or on the internet for the latest pricing.	**TV projectors** are quite different to the other (direct-view) types outlined here. Ideal for use with movie playback, they are at their best in a dedicated viewing room with a powerful surround-sound system. They can produce much bigger pictures than those of other type of home screen, but require viewing in low light conditions. Projectors come in a very wide range of types and prices, discussed further in our Home cinema advice section in the guide. **Advantages** Virtually any size picture can be produced, and some projector types render a picture texture and 'feel' close to that of 35mm film in the cinema. **Disadvantages** Commitment is required – a projector does not easily lend itself to everyday TV viewing, so usually has to be additional to that already in the lounge. Prices can be high and replacement lamps expensive. **Prices** A modest LCD technology projector need cost no more than £250; a good mid-range type with a bright sharp image, say £1500. Above that, the sky is the limit!

Television Viewer's Guide

Curved TV screens

For many years TV screens have been flat and thin. According to the TV manufacturers curved ones are now the latest must-have TV technology. Should you now buy one? We examine the pros and cons.

Curved TV screens are beginning to get some traction in the UK market, albeit at high prices. LG, Panasonic, Samsung and Sony all offer large curved TV screens, with South Korean companies LG and Samsung leading the field. All these manufacturers claim advantages for them over flat ones. Others declare that curved screen televisions are nothing more than a design gimmick, one that will quickly die off once users realize that anyone watching from the periphery has a worse view.

Curved screens can be manufactured in LED or in OLED form. Samsung has even demonstrated a screen with an adjustable curve whereby the press of a button will take it to any point between maximum curvature and flat. Panasonic, Samsung and Sony are currently offering curved LED screens, whereas LG models include OLED, for which you'll pay a premium.

Let's start by looking at some of the claimed advantages for curved TVs.

A wide field of view and a more 'immersive' viewing experience

According to Samsung, 'With a curved screen, your viewing experience reaches another level of immersive realism. It provides a panoramic view that wraps around you and draws you into the picture.' Whilst we'd agree that sitting close to a large curved screen does provide a more immersive experience to a small degree, it does not match the immersive effect of a curved IMAX cinema screen!

There are benefits, but only so long as you're sat in roughly the right position - in the 'sweet spot' in line with the centre of the screen. This is related to the size of the screen, so very large curved screens will have a larger sweet spot. If you've a large family it may not be practical to squeeze everyone into an ideal viewing position. The best viewing is only obtained within an arc of about 50 degrees from the screen centre, more restricted than with a flat screen, so that a curved type embraces – for perfect viewing – a smaller group of people than a flat one.

Those outside the primary viewing area may see geometrical distortion, in the form of image compression, on the side of the picture closest to them because they are viewing it at an acute angle. Also, when watching off-axis you tend to look at what's opposite you and are drawn to the far end of the screen, rather than the middle.

Pictures have more depth and contrast

One thing some people say when they first try out a curved TV is that the picture appears 3D and has more contrast. Samsung claims that their curved screens give you a greater feeling of depth by applying different levels of contrast enhancement on different areas of the screen. We haven't noticed any real 3D effects, but the screens we have seen have offered good contrast when viewed centrally.

It's claimed that the screen's curve can also improve contrast by focusing the light from the screen more directly at viewers. This is difficult to quantify, but we feel that the improvement in contrast is as much about the picture processing applied to the image, and the overall performance of the screen, as it is about the curved design.

Disadvantages

In terms of the disadvantages with the technology, there are a number of key issues, relating primarily to the price of this new

CURVED TV SCREENS 59

LG 55EG960V, 55-inch curved 4K OLED screen - £3799.

technology, reflections picked up by screens, and the practicalities of wall-mounting.

Price
You'll pay more for a curved screen. They cost more to make than flat ones and manufacturers can charge a premium. In time, if curved screens catch on, the difference may disappear as production economies of scale kick in, but for now the price difference could be a reason to stick to flat screen technology.

Screen reflections with curved TVs
One benefit mentioned by supporters of curved screens is that they pick up fewer reflections than flat screens. Whilst this is true, we have noticed that curved screen TVs distort and increase the size of any reflections that they pick up - rather like a mirror at a fun park.

Ideally curved screens should be used in as dark a situation as possible, and positioned facing at right angles to main windows or other likely sources of light.

Wall mounting looks strange
While curved TVs certainly look stylish when mounted on desktop or floor stands, they don't lend themselves to wall mounting. The curved edges do not sit flat on the wall and the TV does not look so neat. It will also protrude further into the room, whether stand-mounted or wall-hung; and this can be a collision risk in some circumstances.

Of late, curved screens have been enthusiastically taken up by most leading manufacturers, but it remains to be seen whether the buying public will be as keen on them. Indeed it has been claimed that the feature has been introduced to give a boost to sales, an incentive to buy in a market which has levelled off lately in the face of the recession and of the virtual rejection of 3D by buyers. Some commentators have suggested that the technology exists because 'manufacturers can' rather than because it offers a worthwhile enhancement of the home-viewing experience, which is very different to that in a large cinema.

If you are a lone viewer, then it might be worth choosing a 60 inch or larger curved TV, but for the average home, we feel that the benefits of a curved screen are questionable and the drawbacks tangible. The additional cost, over a flat screen model, may be better invested in a larger, or better flat screen TV, or a soundbar to improve your TV sound.

Models
The initial leaders in curved screen production

Television Viewer's Guide

CURVED TV SCREENS

were Samsung and LG. The Samsung range includes the HD UE series at £1200 to £1700 and the cheapest curved screen we've found: model UE40JU6740 at £800.

LG are strong on OLED curved screens, including the huge and exotic 77 inch model 77EC980V– at a cool £25000!

Relative newcomers to the curved TV market Panasonic and Sony have a growing range – see below for examples.

Samsung 65JS8500, £2700. 65 inch curved 4K LED Smart TV with two Freesat HD tuners. The Super Smart platform enables 4K streaming via the internet and two-way contact with Samsung smartphones. Active 3D technology is incorporated, along with a Multi-Link screen which can display two pictures simultaneously.

LG 65EG960V, £5000. 65 inch curved 4K OLED Smart TV with excellent black level, including 4 Colour Pixel array. The webOS smart platform enables 4K streaming from internet sources like Netflix and Amazon Prime as well as HD content via Sky movies and NOW TV. 3D viewing uses passive glasses.

Panasonic Viera TX-55CR852B, £1900. 55 inch curved 4K LED Smart TV with My Home Screen smart feature. Both Freeview and Freesat HD tuners are built in, along with Quad-Core Pro processing engine, dual-screen feature, Netflix 4K VOD capability and an active 3D viewing facility. There is also a 65-inch version at £2750.

Sony Bravia KD-55S8005, £1600. 55 inch 4K curved LED Smart TV with Android operating system, Opera full internet browser and Google Cast. Two Freeview tuners are built in to enable simultaneous viewing and recording on a USB hard drive. 3D reproduction uses an active system.

3D television

3D capability is built into mid- and top-range TV screens and projectors. Here we examine its technology, pros and cons.

3D television has not proved the success that was predicted. It part this has been down to the limited range of content, the other key factor is a comfort thing – do we really want to wear 3D goggles while sat on the sofa? Despite the fall from grace of 3D TV, the major television manufacturers are still incorporating the feature in this year's flagship sets. Where previously its inclusion was shouted from the rooftops though, it's now hardly mentioned with the majority of the hype going to new features such as 4K and HDR.

There are two 3D technologies, both compatible with any 3D programme source.

Passive 3D uses simple, cheap viewing glasses and depends on polarisation of light, different for each eye. It gives bright pictures, but definition may be impaired.

Active 3D involves more complex (and more expensive, typically £100 a pair) glasses, powered by a small internal rechargeable battery. They receive wireless signals from an emitter on the TV frame, in response to which they rapidly and sequentially shutter each eye in turn, so that each only receives its own image, alternately displayed on the screen. This gives a dimmer but usually smoother, sharper and better view than a passive system. When deciding between the two it's a good idea to check some 3D footage on the model in prospect.

Picture sources

As with UHD, various internet sites offer 3D material, but most viewing comes from Blu-ray players. Sky shut down its full-time 3D channel in mid 2015, and now offers only an on-demand 3D service, along with Virgin Media – these of course are subscription channels. The availability of 3D on TV sets has also diminished somewhat in the last year.

Blu-ray offers a very good 3D viewing experience – many movies have been shot in the format – though they are far outnumbered by 2D titles, and come at a premium price. Most current and recent Blu-ray players are compatible with 3D except for the very cheapest, and connection to the TV is via an HDMI port. The new Ultra HD 4K Blu-ray players can handle 3D, but only in HD form. In late 2015 UHD 4K Blu-ray discs had yet to reach the market.

Pros and cons

3D can enhance viewing, although the need to wear special glasses (which can fit over prescription ones) is onerous, especially in the general-purpose lounge as opposed to a dedicated home cinema room. For a small minority of people 3D viewing has a flicker effect, or can induce eye fatigue, dizziness or other discomfort. This is more prevalent with active systems. In addition to this between 5 and 10 per cent of viewers are estimated to be afflicted by 'stereo blindness' whereby they cannot perceive the depth dimension of 3D TV images.

The picture dimming of active systems, with each eye receiving less than half the available screen light, is not a major impediment with today's high brightness screens, though the extra weight and cost of active viewing glasses, and the need to recharge their batteries is a disadvantage.

Take-up

The enthusiasm of buyers for 3D TV has not fulfilled the initial hopes and wishes of TV manufacturers due to the need for special glasses and the limited range of content available.

3D has become just another feature of most TVs except for small and inexpensive ones, and a good proportion of viewers do not use the feature. It seems that 3D will remain a back number until it's further developed or content becomes more widely available.

Various manufacturers and broadcasters, including Sony, the BBC and as mentioned above, Sky, have cut or dropped their 3D programming recently.

High-definition television

TV definition, or picture detail, has improved in stages over the years, and there is now a wide choice of equipment and programme sources. Here we describe what's available and what it can do for you.

Three levels of TV definition are in use in the UK. The established SD standard is still widely used. In 2009 it was supplemented by HD on satellite and terrestrial broadcasts. A recent innovation is UHD, Ultra High Definition, also known as 4K TV.

SD – Standard Definition

SD TV offers a very acceptable level of picture detail given a good programme source. The widescreen image contains about 600,000 pixels (picture elements) in a 1024 x 576 formation, and is all that's required on relatively small screens, say up to 30 inch size. SD TVs are cheap to buy and the programmes come free – apart from the cost of the TV licence – from Freeview and, once the set-top box has been purchased and installed, from Freesat and (perhaps more grudgingly!) from Sky in their FreesatfromSky package. Many old movies and TV shows are broadcast in SD anyway.

HD – High Definition

The main alternative to SD is HD whose picture provides over three times as much detail in a 1920 x 1080 (about 2 million pixels) image, and it is in this standard that TV programmes are produced and, on many channels, broadcast, with the SD transmissions being downscaled (see below) from them. HD renders a much sharper and more detailed image than SD, most noticeable on large screens, for which it is really essential for good pictures.

HD broadcasts come in two slightly different standards, set by the broadcaster. 720p pictures are made up of 720 lines per frame, transmitted in sequence – one after the other in real time, thus progressive, as displayed on a computer system. 1080i pictures have more lines per frame, but these are interlaced in successive fields so that two complete fields are required to form a whole picture frame. 720p is better at handling fast motion in news, sport and similar programmes, while 1080i performs better for very fine detail: ideal for documentary, art and nature programmes. In comparison tests 1080i appeared to offer slight advantages over 720p in terms of sharpness, smoothness (less jagged curves and angles) and colour toning. Even so, most viewers can see little difference between these standards.

You'll find a reasonable range of HD content from the main public service channels subscription free on Freesat and Freeview. If you would like more HD content – particularly sport and movies, you should consider a subscription to Sky or Virgin or another pay TV provider.

Large screens can display 1080p, Full HD. It is available from Blu-ray players, but not yet from broadcasters because of its large bandwidth demand.

Ultra HD / 4K

The latest system to become available, UHD, also known as 4K, provides about four times as much detail as HD pictures, with a structure of 3840 x 2160 pixels, about 8 million. UHD picture sources at present are largely confined to internet streaming from Amazon and Netflix, for which a fast broadband speed, say 25 Mbps, is required.

BT TV's Ultra HD channel launched in August 2015, and Sky has announced that it will begin a UHD service in 2016. However, UHD content is not likely to be broadcast on Freeview or Freesat in the foreseeable future due to limited spectrum availability.

Existing and even new UHD TV sets may need updating for UHD reception because the standards had not been finalised in late 2015, and at the very least an HDMI port of v2.0 or higher standard is needed, along with software/firmware upgrades and perhaps new internal hardware. It is important to establish with the dealer and/or manufacturer who will foot the bill for these! Some UHD material

4K AND HIGH-DEFINITION TELEVISION

4K resolution is 4096 x 2160 pixels. In order to fit this higher resolution picture on to a normal 16:9 picture format, it has been altered to 3840 x 2160 – still four times the total number of pixels on a Full HD 1080p screen.

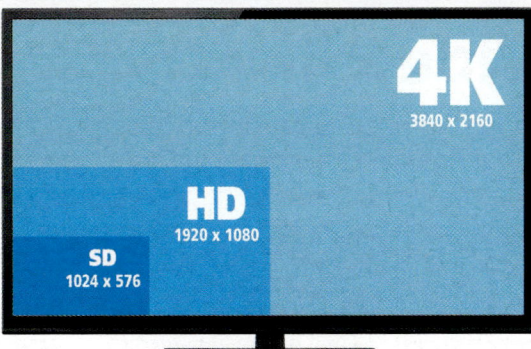

will be broadcast by satellite operators other than Sky. Meanwhile most UHD viewing is of programmes upscaled from HD in the TV or an external box as described below.

Up and downscaling

Most TVs, disc players and set-top boxes have the capability of changing the image structure to match the source picture to the display screen. For small and 'legacy' screens pixels are removed and the result smoothed for presentation. In upscaling new pixels are added, derived from adjacent ones in space and time, to give an apparent increase in definition and crispness – a sort of halfway house between SD and HD or between HD and UHD. Although it's not possible to insert detail that was not there originally the result can be a definite improvement; much depends on the design and performance of the processing engine. All HD and UHD TVs incorporate upscalers.

UHD (4K) TV models

HD operation is standard in all TVs except very small ones, and in Blu-ray players, most of which have SD upscaling (for ordinary DVD discs) built into them; all are capable of working with 1080p when required. UHD TVs are somewhat thinner on the ground, though gradually catching up – industry experts estimate that 1 million UHD TVs were in use at the end of 2015 and predict that they will become the norm in 40 inch and larger screens by the end of 2018.

Meanwhile a good range of UHD TVs from the major manufacturers are available now. Prices have dropped: UHD screens, averaging £3000 in early 2015, can now be bought, albeit in a small 40 inch size, for as little as £450 (Hisense brand), while £1000 will now secure a respectable UHD TV. **Panasonic model TX-50CX802** is an LED UHD 50 inch model retailing at about £1500; 40, 55 and 65 inch versions are available at proportionate prices. Three HDMI v2.2 ports are provided, also three USB sockets and Scart and component sockets for legacy equipment. This Smart model provides very good pictures.

Sony has an excellent Smart TV model: type KD55X8509C at around £1300. For this you get a sharp and rich picture thanks to its Triluminous (Quantum Dot) screen. Using the Android platform, web streaming and programme catch-up is easy; Google Cast is also provided here. Ethernet and WiFi are featured, and there are two Freeview tuners.

Model 55EF950V from LG comes towards the top of the UHD TV range, standing out from those described above in terms of picture quality, and particularly black level rendering, because of its OLED screen. It comes at a premium though: the price is around £3500. Other features are similar to those of the LED TVs listed above.

8K – super Hi-Vision

8K TV goes much further than 4K technology, with 7860 x 4320 pixels. This is probably a step beyond what most people will ever want because you really need a screen the size of your whole living-room wall to do it justice! Broadcaster NHK plans to launch 8K in Japan in 2020. Sharp launched the world's first 8K TV in 2015. Costing around £86,000, interest is expected to come from broadcasters and other companies testing the format.

Video discs

In the nineteen years they have been available, video discs have enjoyed great popularity and have included popular formats such as DVD and Blu-ray. We look at your buying choices.

DVDs and Blu-ray discs contain compressed digital video and sound data in the form of a microscopic spiral of pits embossed in the surface or – on home-recorded oncs – a spiral of tiny burn marks. They rotate at high speed while being scanned by a low-power laser whose beam reflection provides programme and housekeeping (guidance, speed etc.) data.

There are three basic types of player: **The original DVD format**, using a red laser, is capable of SD (standard definition) reproduction and programme time of about two hours. It has been largely superseded now by Blu-ray, and is not recommended for purchase, even though its price is low, ranging through £20 – £55.
Blu-ray cost from £50 upwards. They are compatible with ordinary DVDs, and use a blue laser with a smaller light spot to increase the data capacity on the disc. This opens the way to high picture definition, up to 1080p, to match the capabilities of current TVs and broadcasts. A huge range of movies is available on Blu-ray discs, though they are more expensive than DVDs. Both DVD and Blu-ray players can upscale ordinary DVD pictures to increase perceived definition provided they are connected via HDMI.
The new Blu-ray UHD/4K is the third type. With 3840 x 2160 pixel picture definition, it is still compatible with existing Blu-ray discs, and features other advanced technologies in terms of colour gamut, fast picture refresh and dynamic range of the image. It calls for a UHD-compatible TV with very large screen and HDMI v2.0 (or higher) connections.

Connections
Current disc players no longer feature analogue/Scart connectors, the norm being HDMI sockets, typically just one on the cheapest types. If your TV has no HDMI port it is very old, and in need of replacing!

Upscaling is very common, in which picture detail appears to be enhanced, but this only duplicates a function of the TV, and may not do as well as the latter, especially with a cheap player and a high-quality TV.

Blu-ray discs can provide sound in many forms, from plain stereo to multi-channel surround and Dolby Atmos soundstages, so if you possess – or aspire to – any audio add-on from a soundbar upwards it's wise to choose a player with more than one HDMI port. Most TVs automatically switch to the disc player when it's turned on or into play so that operation is easy.

Features
Even budget Blu-ray players can offer web access, particularly useful if your TV does not offer this facility. This requires an internet connection, the faster the better. Consider a Blu-ray player of the same make as your TV, where a single remote control can generally be used to drive both.

Blu-ray web portals generally mimic those of Smart TVs, with apps for on-line content, typically for broadcasters' catch-up services; and Facebook, Twitter, Netflix, Amazon, YouTube etc. Some Blu-ray players can access the Ultraviolet digital locker, a bridge between physical disc media and the burgeoning world of streaming on line.

Another feature of many Blu-ray players is Miracast wireless screen mirroring. This facilitates viewing pictures on your TV from an Android smartphone or Windows computer. Many new Blu-ray players can integrate with smartphones, tablets and PCs in this and other ways.

Also featured in some models are gaming, Ethernet coupling and WiFi streaming. Even if you do not presently require all or any of these bells and whistles most add relatively little to the player's cost, and come as standard with mid-and high-range players whose other benefits may well appeal to you.

A useful category of Blu-ray equipment

VIDEO DISCS | 65

Pioneer BDP-LX58, £600. Excellent picture and sound performance as well as Ultra HD 4K upscaling, as well as an Ethernet socket to enable connection to your local network.

is the All-In-One type. These combine a disc player with a Dolby Digital multichannel audio amplifier in one box which can conveniently sit below the TV, needing only to be connected to it and the supplied loudspeakers to create a simple to install home cinema system. Ranging from soundbar to 7.1 types, they start below £200. Blu-ray decks generally support 3D movie discs, many of which are available for use with a suitable TV.

Disc recorders

If you're into archiving TV or camcorder footage the concept of burning your own disc may appeal: the recordings made in '+' receiving boxes are not removable or portable. The benefit, however, is mitigated by the fact that Blu-ray recordings can be blocked by copyright protection. Disc recorders include TV tuners for Freeview reception with programmable timers. They are more expensive than 'straight' players, and usually offer smart and other features.

Panasonic model DMR-EX97EB, £350, incorporates UHD upscaling, two Freeview HD tuners and a separate 500GB HDD for broadcast recording, while £430 model DMR-BWT740EB also records on Blu-ray and HDD, the latter a 1TB device: up to 518 hours of SD and 259 hours of HD recording can be made. This model has a range of smart features like catch-up TV, DVD upscaling and built-in WiFi.

Video discs can also be burned using a suitably-equipped home computer.

Regional coding

Beware of buying video discs intended for use in other world regions – they may not play on your deck. DVDs for UK use are coded Region 2, and Blu-ray ones Region B. Foreign vendors do not always make this clear!

Players to buy

Starting near the low end, the Sony £100 model BDP-S5500 is a very small (230cm wide) Blu-ray player with a single HDMI connector, Ethernet port, co-ax digital audio output, 1080p upscaling and 3D capability. It quickly boots up and gives good pictures and sound. Similar models at about the same price come from LG, Panasonic and Samsung.

Samsung BD-J7500, £170, offers twin HDMI ports and comprehensive sound connectivity in a good-looking package at about £170. UHD upscaling, video streaming, smart operation with catch-up TV and smartphone linkage are featured, together with dual-band (2.4/5GHz) WiFi.

Further up the scale the Pioneer BDP-LX58 at £600, pictured above, is more suited to home cinema use because of its excellent picture and sound performance. It's well equipped in the connectivity and features stakes with two HDMI and two USB slots, co-ax and phono sound outlets and a LAN socket for access to its smart apps. 4K upscaling and 3D playback are also featured.

The world's first UHD Blu-ray player, model UBS-K8500, was revealed by Samsung in late 2015. Expected to be available in Europe in Spring 2016 at a price (unconfirmed) around £450, it uses dual-layer 66GB or triple-layer 100GB discs offering data rates of over 100Mb per second, way beyond what's available from web-based UHD picture sources.

The future for discs

Video discs are hugely popular, can be inexpensive, and offer a huge range of movies. The advent of UHD/4K will give them a boost, but in the medium and long term their future is threatened by internet streaming services, especially as online data speeds increase and as Smart TVs proliferate.

Television Viewer's Guide

Home cinema and TV sound

Home cinema, also called home theatre, seeks to reproduce the feel of a cinema in your own home. In recent years the advent of high-definition video, larger TV screens and Blu-ray have all helped provide a higher quality picture. Although modern slim HD televisions provide pictures that look fantastic, they usually offer disappointing audio. With a home cinema sound system you can enjoy top quality audio, clear dialogue and realistic surround sound effects.

Watching movies from the comfort of an armchair at home is easy these days, with a huge range of films available and excellent equipment to bring them to life.

Home cinema can be as simple as a large flat screen TV and a few extra speakers in the living room, or as advanced as you want to make it. Whatever form it takes, the basics of a home cinema system are the same: a room to watch in; a large screen; a surround sound system, ideally with a subwoofer to give extra bass; and a source of film or programme.

Although some home cinema enthusiasts dedicate a room to their set-up, there is no need to go to that length. With a simple set-up and a little thought you can dramatically improve the quality of your TV picture and sound.

Choice of TV or projector

With the availability of reasonably priced large screen televisions, the emphasis of most home cinema set-ups is now about improvement of the audio quality. That's not to say that projectors do not have a role to play when you want a really large screen.

Television

For a truly immersive cinematic experience how big a TV should you choose? For most people the answer will be as big as they can afford. Large screen prices have dropped considerably in recent years, with 50 inch screens available for under £350.

You should also consider the recommended ratio between screen size and seating distance. This is usually given to be between 2 and 2.5 times the screen size; although with 1080p TVs you will probably have to sit closer than 1.5 times the screen diagonal to be able to discern the detail.

Most people however don't choose to sit at the optimum distance from their screen. Instead a range of factors such as room layout and what distance feels comfortable to them determines this. Most people sit about 9 feet from their TV. At this distance a TV screen size of around 55 inches is recommended. The bottom line however, is that screen size is one of personal preference. One useful tip is to make a cardboard cut out of the same size as the TV you are planning to buy. Position it in your room to see how well it fits.

Projectors

Projectors are the cheapest way of producing pictures larger than 65 inches, and they offer the closest approach to the real cinema experience. They are not as simple to set up as a TV-based home cinema system, however, and you will also need a screen.

We have listed the key types in the panel opposite. The biggest issue with any type of projector is image contrast - it's vital to eliminate as much ambient light as possible. Too much light will wash out the image. 4K/Ultra HD projectors are available – at a price!

Viewing sources

The most prolific film platform is Sky, with dozens of dedicated channels in both standard and high-definition. Other platforms such as Freeview, Freesat and Virgin carry high-definition content too, but currently for the highest widely available quality, 1080p, Blu-ray is the best choice, particularly if watching movies is your aim.

HOME CINEMA AND TV SOUND SYSTEMS | 67

Sorting out the sound
With the improvement in picture quality that HD and larger screens have brought, the most important thing to sort out now is your TV's sound. With many modern flat screen TVs the speakers are squeezed into limited space, often facing backwards.

A home cinema sound system solves this problem by positioning several speakers around your room. It then decodes the surround sound information stored on DVDs, Blu-ray discs and transmitted by some TV programmes to deliver an immersive sound that makes watching films and TV far more enjoyable.

Let's look at the options. There are three main types of sound system available: all-in-one systems, separates and soundbars.

All-in-one systems
All-in-one systems are the easiest way to arrange a home cinema system. They contain everything you need to get started; the amplifier, speakers, cables and instructions on how to connect it all. A Blu-ray player may also be included.

Equipment is available as 2.1, 3.1, 5.1 and 7.1 systems. This refers to the number of speakers included in the set-up. Simple 2.1 home cinema systems can cost less than £100, but for around £50 more you can get a reasonable 5.1 system that provides true surround sound.

Separates
You can also buy the components of a home cinema sound system separately – amplifier, loudspeakers, cables and a Blu-ray player. This route may appeal to the committed audiophile or home cinema enthusiast wanting to perfect their set-up. Buying this way also makes it easier to upgrade different components over time.

Buying separates is, however, more expensive. A basic home cinema amplifier will cost around £200. Throw in cables, speakers and a disc player, and you're looking at, perhaps, £350 for a separates system. If your budget allows, it's easy to spend thousands of pounds on a system built from separate components.

Projector types and screens
Three main types of technology are currently used by front projectors: DLP, LCD and LCOS.

DLP (Digital Light Processing) uses
a small electromechanical chip, the size of a postage stamp, with hundreds of thousands of tiny vibrating mirrors. Colour is added by a rotating RGB filter wheel, or by providing separate DLP chips for each primary colour: with these a possible 'rainbow' effect in the image is avoided. Some later model DLP projectors don't have a hot lamp; instead they use a brilliant triple-LED light source, very rapidly sequencing between red, green and blue, eliminating rainbow artefacts and having a better service life. Prices of SD DLP projectors start at around £300.

LCD (Liquid Crystal Display)
projectors use similar technology to that used in flat-panel TVs. They tend to be small and light, with good colour reproduction, but a wide price/performance range. Cheaper LCD-types can suffer from poor contrast and brightness, and show a 'chicken-wire' effect, whereby individual pixels can be discerned within the image. Prices range from £300 for a basic model to £3000 for a highly-specified HD model.

LCOS (Liquid Crystal on Silicon)
These can be thought of as a hybrid of DLP and LCD, having a liquid crystal layer on top of a mirrored surface. Black levels and contrast ratios are both excellent. JVC projectors (using their version of LCOS, called D-ILA) have well-rated contrast ratios. Sony's version, called SXRD, is comparable. Prices range from mid-range to very high, with Sony's 4K projector VPL-VW1100 costing £18000.

Screens
Don't forget your screen. They range in price from £100 to many thousands. They come in all shapes and sizes – electric, pull-down and portable. We don't recommend just using a painted wall as any sort of texture will probably be visible.

Television Viewer's Guide

68 HOME CINEMA AND TV SOUND SYSTEMS

A simple 2.1 system

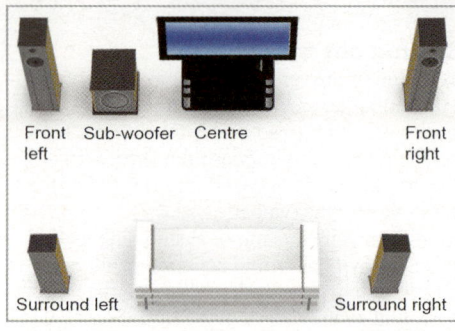

A 5.1 system the addition of two rear speakers gives a much better surround sound effect.

Soundbars

Soundbars comprise two or more loudspeakers in a bar-shaped box. Sometimes they are in the form of a plinth for the TV to stand on; others go on the floor in front of the screen, or can be wall-mounted behind it. They contain a mains-powered amplifier and control system and need only a single cable to connect to the TV; some have Bluetooth WiFi. They typically include a subwoofer (bass loudspeaker), sometimes integrated into the bar, but better provided as a separate unit.

The sound to each speaker is electronically processed in order to create a 'virtual' surround-sound effect, despite their lack of rear speakers. Models are designed to mimic 2.1, 5.1 and even 7.1 systems.

We haven't found their audio quality as good as that of all-in-one systems.

Prices start from around £50 for basic dual-speaker soundbars. Products from the mainstream audio-visual manufacturers such as Sony, Samsung and Panasonic are typically priced from £120 upwards.

How many speakers do you need?

If you are purchasing an all-in-one system, or building your home cinema system from separates, how many speakers do you need?

2.1 Systems

The simplest home cinema sound systems have two front speakers, one each side of the TV set, and a sub-woofer (the '.1' of the 2.1).

The sub-woofer is the speaker that delivers rich, deep bass, adding drama and realism to the movie. It can be positioned wherever convenient as the bass sound produced is non-directional.

You won't get true surround sound with

Philips HTL 5140 A good compromise between performance and cost, this product features a total output of 320W, HDMI hookup, wireless subwoofer and Bluetooth streaming. It costs about £240.

HOME CINEMA AND TV SOUND SYSTEMS

these systems, but you should experience a significant improvement over the speakers built into your TV.

5.1 systems
These deliver full surround sound using three front speakers, two rear speakers, a subwoofer and often include a Blu-ray player. The addition of a centre speaker at the front enhances speech, and two rear surround speakers, left and right, help to create a far more cinematic sound experience. Unless you purchase a system with wireless rear speakers, you will need to run cables to the rear speakers. Even if wireless they will need to be connected to mains power.

7.1 systems
With an additional two rear speakers, 7.1 systems are more challenging to configure and are for the home cinema enthusiast.

Dolby Atmos
The latest innovation in home cinema sound is Atmos by Dolby. It adds two or four ceiling-mounted or upward-firing speakers to a 5.1 or 7.1 system for an enhanced special sound stage, along with a new 'tighter' (object-based) coding system. A new surround processor is required, generally incorporated in the amplifier box, though some existing late ones are amenable to software updates. Atmos Blu-ray discs are becoming more widely available in 2016. All discs and systems are inter-compatible.

Cabled or wireless
With a 2.1 system you will not usually need to run cables too far. However, with a 5.1 system the rear speakers will need connecting to your system's amplifier. Consider a system with wireless rear speakers.

Don't spend too much on cables
There's no need to spend large sums of money on premium-priced cables. Poorly made budget cables are not a good buy either, but most tests have shown that, at best, listeners can only detect a very slight improvement with more expensive hi-fi cables.

With HDMI we have not found any difference in quality between basic and expensive cables, at least for short runs.

Home cinema tips

Set up your TV
Television manufacturers generally set up their screens before they leave the factory for shop displays. Ditch the 'Dynamic' preset mode that televisions are frequently set to. It gives a harsh, sharp and over-bright image, usually unsuitable for conditions in the living room or home cinema. Modern TVs have many settings to enable you to get the best balance of sharpness, contrast, brightness and colour.

Positioning your screen
To avoid a crick in your neck, don't position your TV higher than around 15 degrees above your eye height when viewing. Above the mantlepiece is not always a good screen position!

Minimize over-bright backlighting and reflections by positioning the screen, whether wall or stand-mounted, at 90 degrees to the room's main window.

Darken your room
Contrast is one of the most important aspects of picture quality. It is particularly important to darken the room when viewing projected images. However sitting in a dark room with the lights out can cause eyestrain, particularly if you are viewing a bright television display. Philips' Ambilight system can contribute to comfortable viewing and improve perceived picture quality. It illuminates the wall behind and around the screen, or the TV bezel, with soft light, changing colour in sympathy with the picture. Alternatively put a dim lamp behind the TV!

Sound tips
For that home cinema feel there is no better upgrade you can make than upgrading your TV's sound, preferably with a 5.1 system with rear surround sound speakers.

Curtains and carpets
Hard laminate floors and large windows without curtains are not good for surround sound reproduction.

Get expert advice
If you are new to home cinema it's a good idea to talk to an expert before you purchase expensive equipment.

Television Viewer's Guide

Connecting your TV equipment

Now that we have digiboxes, video disc players and also internet connections, the wires, plugs and sockets have proliferated. Even so, it's not too difficult to link all the boxes together so long as you take a logical approach and follow the instructions in the user handbooks. We start with standard-definition and analogue equipment.

Scart connector
Analogue TV signals and stereo sound are carried in a Scart lead, which has 21 connector points. It conveys an analogue composite signal in which all the video information passes (in relatively low-definition form) along a single conductor to the TV. This mode is now obsolescent, but continues in use with older (non-HDMI) TV and video equipment.

Better definition and picture quality are provided in a Scart cable by RGB signals, in which the picture's primary colour ingredients, red, green and blue, each have their own conductors in the cable. Although still analogue these are derived from digital picture datastreams in set-top boxes and video disc players, and give reasonable results in standard-definition (SD) on small and medium sized screens.

Analogue stereo sound is carried in Scart links with quite good quality, but surround sound is poorly served: only a crude analogue form of it, Dolby Pro-Logic, can pass through. A Scart link also performs a couple of control functions in switching the set to its own signal when 'on' or 'play' is selected at the source; and in auto-setting the correct aspect ratio or picture shape.

Some large-screen TVs are still equipped for RGB operation from their Scart sockets. Scart leads can also convey **S-Video** signals, primarily from old S-VHS decks and camcorders, so long as the TV has been menu-configured for it. S-video also has its own dedicated connector (four-pin S type). It's used in conjunction with a twin-phono lead carrying stereo sound. A fifth method of conveying pictures, used mainly with DVD players, is **component video**, carried in three phono cables.

Cable quality
Cheap Scart types are a false economy. They can cause picture and sound disturbance due to external interference and crosstalk between the signal conductors. Cables with individually-screened conductors are a good investment. They generally have gold-plated contacts for reliable signal transfer.

Basic Principles
- Use a good quality Scart cable to avoid the problems described above; the longer the cable the more important this becomes.
- Use digital (HDMI) links when possible; they give better quality than analogue ones.
- Read the manufacturer's instructions.
- Once you have the system working draw a diagram or take photos to show how the cables run - useful should you ever need to disturb it.

Older equipment - SD connections

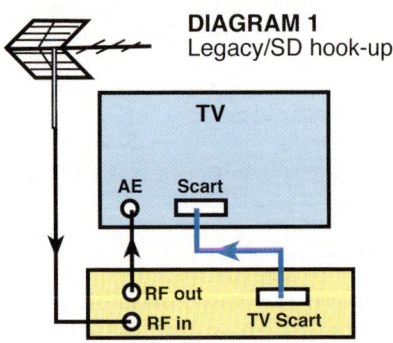

DIAGRAM 1
Legacy/SD hook-up

Diagram 1 shows a simple hook-up between a TV set and a recorder or disc player. In the case of a video deck or disc recorder the aerial signal has to loop through it so that simultaneous viewing and recording of different channels is possible.

CONNECTING YOUR TV EQUIPMENT | 71

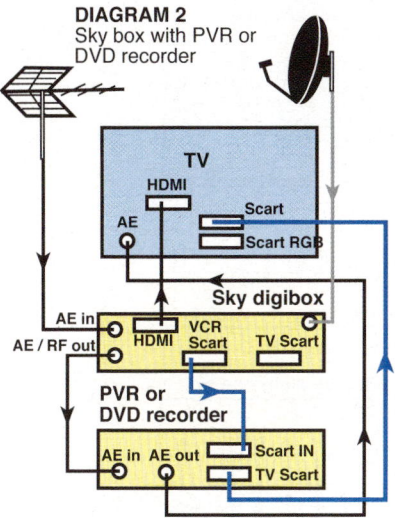

Diagram 3 shows a common set-up, a Freeview box/PVR and a DVD player. In this example the TV has two RGB-compatible Scarts and the DVD player does not have a RGB Scart loop-through. If the DVD player and TV have HDMI connections use these in preference to Scarts.

Some older Freeview boxes have only one Scart socket and an S-video socket. The latter can supply a video signal almost as good as that from an RGB Scart.

Diagram 4 shows an alternative set-up involving a TV, digibox (satellite, Freeview or cable, which may sport an internal HDD recorder, i.e. a DVR) and a 'legacy' (no tuner fitted) DVD or Blu-ray deck, perhaps with recording facility. Both the digibox and the disc player should be set up for RGB operation. The digibox output is routed through the video disc deck to enable recordings to be made off-air to disc – assuming it's a recorder and not just a player. Late and current video disc recorders have internal tuners, calling for their own aerial loop-through.

Diagram 2 shows a Sky digibox integrated into the system. Connect the Sky box directly to the TV using an HDMI cable whenever possible. Both the TV and the digibox should be configured for RGB if you have older equipment without HDMI connections.

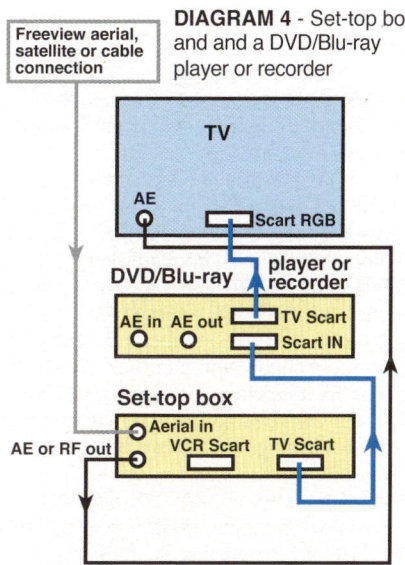

Television Viewer's Guide

72 CONNECTING YOUR TV EQUIPMENT

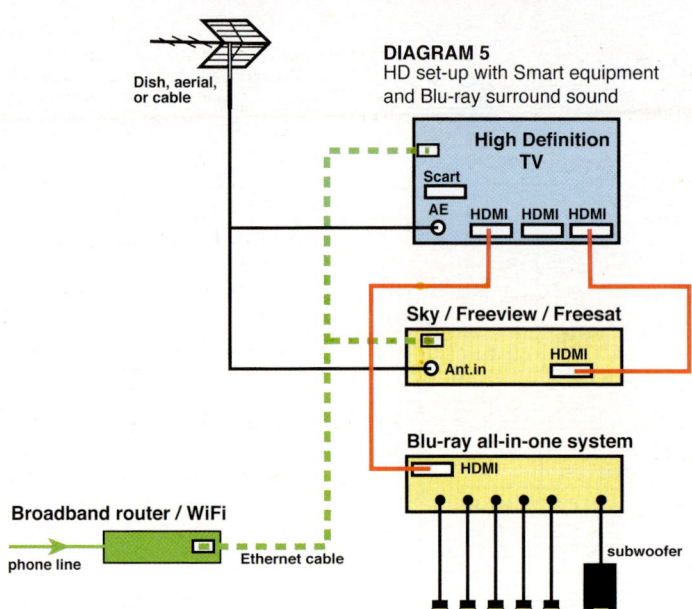

DIAGRAM 5
HD set-up with Smart equipment and Blu-ray surround sound

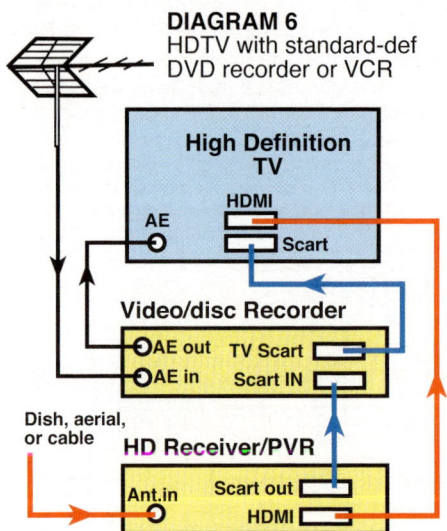

DIAGRAM 6
HDTV with standard-def DVD recorder or VCR

High-definition equipment
All LCD/LED and OLED TV sets are fitted with at least two digital HDMI input ports. High-definition is available from disc players, games consoles, Freesat, Freeview and the internet. All HD boxes are fitted with HDMI ports for the best possible signal transfer.

Diagram 5 above, incorporates an all-in-one DVD or Blu-ray player/surround-sound outfit and a digital set-top box, all linked up with HDMI cables to a modern Smart TV set. Freeview or Freesat transmissions provide a subscription-free source of standard and high-definition programmes, hooked straight into the TV's aerial socket. The TV, via its Smart interface, can access the internet, as can some set-top boxes, while modern equipment may have wireless connection facilities. All these diagrams are adaptable – you can mix and match your own equipment following parts of any of the wiring guides shown here.

Diagram 6 combines standard-definition video recorder equipment with a relatively simple HDTV set-up using Scart and HDMI. The Scart link from the HD set-top box to the video recorder enables recordings in standard-definition.

HDMI systems incorporate a form of security handshake (HDCP) to combat pirate copying, whereby no transfer can take place without an assurance that the data receiver is purely a display device, incapable of recording. The

Television Viewer's Guide

CONNECTING YOUR TV EQUIPMENT 73

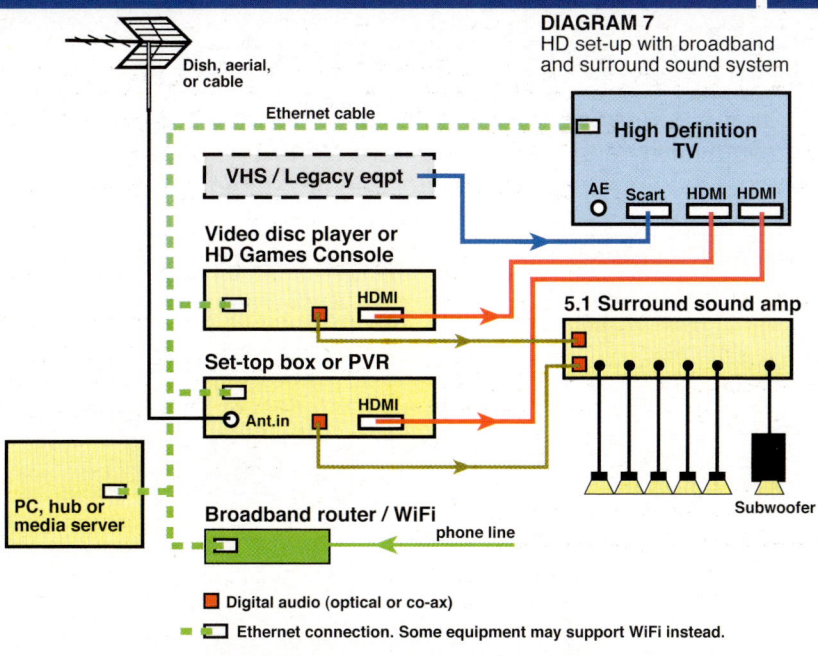

latest HDMI protocols support advanced forms of digital surround sound, smoother image-colour reproduction and 3D.

Diagram 7 shows an example of a set-up with high-definition TV, gaming and movies, plus Dolby 5.1 surround sound. Here the Blu-ray player could have a separate surround sound amplifier as shown, or as is increasingly found, it could be built-in. Access to the internet for a Smart TV and 'connected' set-top boxes and games machines could be 'wired' using Ethernet cables, or if the equipment supports it, via WiFi.

HDMI practicalities
HDMI cables come in lengths from one metre to 20 metres. We have not found any performance advantage to be found with expensive HDMI leads.

Current TV screens boast three or four HDMI ports. Where there are more cables than TV sockets, selector switches are available. Links between DVI cables and HDMI ports can be made with adapter cables.

TVs and auto-switch boxes can select HDMI inputs by remote control or by 'here-I-am' electrical flags raised by source boxes when they are switched on or into play mode.

System control via HDMI
We have seen that the obsolescent Scart connection system has two control functions: sound and video signal routing and picture-shape control. HDMI networks also have a system control facility called CEC (Consumer Electronics Control) protocol. This can perform these functions and several more, such as co-ordinated switching in/out of standby mode, timer operations, onscreen display co-ordination, remote control unification etc. The basic operating language for this is common to all equipment makes and models, so that everything in the HDMI loop should be in harmony. If not, one or more of the boxes may need a software update: contact the dealer or manufacturer.

Further help
Hook-up and connection diagrams are often included in equipment manufacturers' instruction books; experts can often be found in Yellow Pages, local papers or directories.

Television Viewer's Guide

Cables and connectors

Poor quality cables are less well screened, pick up more interference, suffer poor conductivity and are likely to have contacts that oxidise. Although cable quality is important, you don't need to spend a fortune!

SCART

Scarts have 21 pins and are often used to connect DVDs, VCRs and digiboxes to a TV and to each other. They carry signals in various forms - composite video gives basic quality; RGB signals offer excellent quality. They also carry a trigger signal which informs the TV when an AV product to which it is connected is turned on.

RF COAXIAL

Radio Frequency (RF) coaxial cable is used to carry an aerial signal to the TV, a set-top box or a VCR. It is particularly important to use high quality double-screened cable with satellite and Freeview installations.

F-CONNECTOR

Used for satellite downleads, this cable plugs into the back of a set-top box or TV. It carries programmes downwards and LNB commands upwards.

S-VIDEO

Better quality than composite video over a Scart, but not as good as RGB Scart. It separates picture and colour information for better images. Can be carried on S-Video compatible Scarts or 4 pin S-Video connectors above. Needs separate audio links.

BNC

Used with some equipment and applications with a composite video signal. It requires separate audio links.

PHONO OR RCA

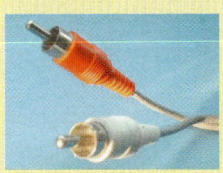

Simple red and white phono leads commonly used with hi-fi systems to carry analogue stereo sound. Also used to carry surround sound (co-axial digital) from video equipment to a Dolby Digital decoder.

Television Viewer's Guide

CABLES AND CONNECTORS 75

COMPONENT VIDEO

A high-end video connection found in some flat screen TVs and projectors. The signal is split into its component parts. Capable of handling a non-HDCP high-definition signal.

HDMI

High-Definition Multimedia Interface – A digital connection system. HDMI carries vision, multi-channel sound and control data. Version 1.4 can carry 3D TV; version 2.0 is necessary for 4K / Ultra HD. The Mini HDMI version is shown below.

USB standard-sized

A common link found in the computer world and in TV equipment.
USB v3.0 is over 10 x faster than USB v2.0.

OPTICAL DIGITAL. Carries a digital sound signal, on a fibre-optic cable, between two digital products - for example from a DVD player, carrying Dolby Digital 5.1, to a digital decoder/amplifier.

ETHERNET

A computer networking system using RJ-45 connection plugs as shown here. Ethernet is commonly used to link PCs, printers and broadband routers. Increasingly it is used to connect Smart TVs and internet connected set-top boxes to a home broadband network.

USB mini and micro

Mini and micro USB types are illustrated above.

Television Viewer's Guide

TV calibration

Television manufacturers set up their screens before they leave the factory to suit the ambience of the dealer's shop window or showroom. This gives a harsh and over-bright image, quite unsuitable for conditions in the living room or home cinema, so readjustment is required.

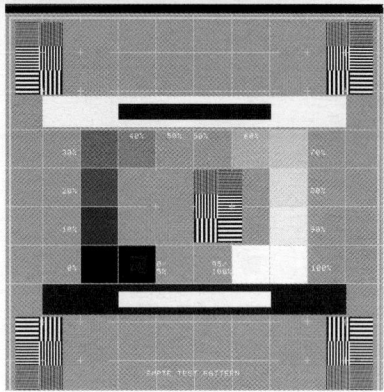

Black and white settings testcard

You can achieve a great improvement in the picture by simple adjustment of the ordinary user-controls in the set-up menu for your television.

Read the relevant part of its instruction book, then set the TV to 'normal' display mode, deselecting options like Active Vision, Dynamic, Game, Live Colour, Sport, Turbo, Vivid etc, and any incorporating the word 'motion'.

Sort out the black and white first

Turn down the colour all the way to zero and set the backlight control (if fitted) to about 50%; and the brightness, contrast, sharpness and tint or hue controls to 50%, halfway or a little more. Darken the viewing room, then – with a studio programme selected or a video disc running (ideally one with a test pattern), adjust the brightness so that the black parts of the picture, e.g. letterbox movie margins, are just disappearing. Do this at the lowest practical setting of any backlight control provided: this gives the best semblance of black, saves energy, and prolongs the TV's life. Now reset the contrast control for the most comfortable glare-free image highlights at the ambient light level (the darker the better) which is normal in the viewing area. If necessary advance the backlight and reset the brightness control as above to achieve this.

Next the colour

The next step is to adjust the hue or tint setting to give the greys and whites a neutral tone, neither cold nor warm in appearance. Next, looking at fine detail or sharp edges in the picture, or narrow gratings in a test pattern, set the sharpness preset for best definition without the appearance of ghosting, 'grass' or flying-dust effects at the vertical black-white-black transitions. Finally it's time to add the colour. Turn it up to the point where human flesh tones, in a studio or test pattern setting, appear normal and healthy. Skin is the most subtle and easily assessed colour hue, and when it looks right other colours are accurately reproduced.

Calorimeter use

Several kits for basic home TV calibration are available. These usually consist of a screen-mounted calorimeter (colour/brightness measurer), PC software and in some cases a DVD/Blu-ray disc containing test patterns. They start at about £90, and can improve picture reproduction in TVs and monitors which use numerical values in their calibration settings.

The main manufacturers of these are Datacolor, Crisp Digital, LaCie and X-rite. In TVs where these presets are separately available for each picture source/input port, adjust them to achieve optimum performance with the source (disc player, set-top box, etc.) normally used, and a trusted or favourite programme or disc running. Still and slow-moving pictures should now be excellent, and you can go on to experiment with motion-correction features to get the best reproduction (least judder, blur, ball-ghosting, 'jaggies' etc.) on fast motion. The game settings offered by some TVs minimise propagation delays in the image processor to help with games requiring fast player reactions.

TELEVISION CALIBRATION

Test patterns are built in to some TV sets; alternatively download from the internet or purchase a set-up disc.

Test pattern sources
Some TV models have built-in pattern generators, while THX-certified movie discs feature a basic set-up tool called THX Optimiser, found in their installation menus. With Sony Blu-ray movie discs, type into the title screen, via the remote control, 7669 then press ENTER to bring up screen-calibration patterns. For £10-£25 you can buy a professional test pattern disc like JKP's Digital Video Essentials to greatly assist the setting-up process. Alternatively, if you are good with computers you can download test images from various websites and burn them onto your own disc.

With most tuners you can find a (low-grade) test card on Freeview channel 200, BBC Red Button. When the welcome screen shows press the yellow key, then select a different TV channel. Now go back to channel 200, wait for 'welcome' then press green. Now follow the on-screen instruction, i.e. press green again for the test card. On some sets and tuners it may be necessary to key in 33582, red, green, yellow, blue after the first press of the green button above, then wait for up to 30 seconds to see the card.

Professional calibration
For about £200-£300 your screen can be adjusted by a professional calibrator, certified by standards authorities ISF or THX. They align the user controls already described with much greater accuracy, then use screen-mounted colour and light-sensors in conjunction with a portable computer running special software to preset many aspects of the picture display. The colour temperature (white/grey tone) is calibrated, then the basic primary colours red, green and blue are set to the correct points, and their intensities matched throughout the brightness range to achieve truly neutral greys at all levels.

The gamma setting is checked and adjusted, ensuring that brightness graduations are properly reproduced on screen: detail is then clearly visible at both ends of the brightness/contrast spectrum. Colour saturation is precisely checked and set to optimum along with other aspects of colour reproduction. All these settings are accessed and adjusted in the factory/service set-up registers, not available to the user, and then permanently stored in the TV's internal memory: you can subsequently change these settings, but it's easy to revert to the calibration levels. Some expensive screens have a direct link for data from the calibrating computer, reducing the set-up time.

Is it worthwhile?
DIY screen adjustment is certainly worthwhile if you want to obtain the best picture. Many people who are really serious about home cinema, wanting to watch movies, sports and documentaries as the producer and director intended, will feel that the cost of a professional calibration is justified. However, for most viewers simply following the above DIY steps will provide a great improvement in picture quality and add to the enjoyment of watching your TV.

Sending video signals around the house

Home networking and wireless streaming are now common. We examine these, along with less sophisticated and cheaper ways of distributing SD and HD pictures.

Freeview
Freeview signals are relatively easy to pipe around the house. On the aerial mast, under the eaves or behind the TV you can install a distribution amplifier, with one input and as many outputs as required, routed, inside or outside, to the secondary TVs. It's powered from an indoor mains unit via one of its output leads. If the Freeview signal is very strong (due to a large aerial and/or a nearby transmitter) and there are not too many branches it may be possible to use a passive splitter instead of a distribution amplifier; it's cheaper and requires no power supply.

Satellite - Sky and Freesat
Satellite signal distribution of Sky and Freesat signals is a little more involved. Each receiver (PVR type) needs its own pair of twin cables all the way to the LNB at the dish. For this purpose LNBs are available with up to eight (Octo type) outputs to cater for up to four boxes. With Sky you'll need a Multi-Room subscription, and any phone connections must be to the same line.

Point to point coupling
Rather than broadcast signals, locally generated AV ones are often required to pass between screens and 'boxes' within the home, e.g. to and from recorders, players, receivers, PCs and TVs. There are several ways of achieving this, with and without cables, and we shall concentrate on them throughout the rest of this section. If your equipment is more than three to four years old (the vast majority of existing gear) it will be fitted with analogue tuners and Scart sockets. For these, so long as you are content with SD (standard definition) pictures and single-channel (mono) sound – reasonably tolerable on small screens in secondary rooms – there are two inexpensive choices: modulators and video senders.

Modulators
Fitted to older set-top boxes and recorders, modulators mimic a, now obsolete, analogue TV signal, sent out on a carrier in the UHF TV band. This can be tuned in on a TV connected to it by cable. Modulators can be purchased at prices from £40, and connect to the source equipment by a Scart plug, and to the receiver via its aerial socket. They are available from suppliers such as Keene Electronics and Satcure.

Video senders
Also working with SD analogue signals, the cheapest wireless video senders typically cost from about £35. They usually have infra-red remote control 'talkback' facility incorporated, and offer a transmitting range, in average conditions, of about 12 metres, depending on the building structure: steel frames and thick walls inhibit the signal somewhat.

HD links
Moving up to HD operation with HDMI connectors, there is a choice of several wireless senders at prices from £75 upwards from manufacturers like AEI Digisender, Triax and Marmitek. Such equipment,

The **Antiference Magic Eye** allows control of a Sky box from another room.

SENDING VIDEO SIGNALS AROUND THE HOUSE

however, has been overtaken in recent years by the growing popularity of WiFi. HD pictures can also be coupled or distributed by wire, for which various kits are available, mainly using co-ax or Ethernet cable.

Sky Remote Eye
Where the output of a satellite receiver is coupled by cable to a secondary TV, typically in the living room and bedroom respectively, the receiver box in another room can be controlled by using the Sky Remote Eye, £5, which relays the control signal via the cable.

Powerline systems
A half-way-house between wireless senders and cable links is the powerline distributor. This plugs at each end into a 13A mains power socket and uses the existing ring-main wires to carry the AV signal. It can be subject to interference, particularly from other WiFi equipment and electric motors. WiFi and Ethernet powerline extenders are available from about £40 per pair, the price dependent on their speed, number of Ethernet ports etc.

WiFi
WiFi devices are simpler to install and connect than those using cables, and are popular and widespread as a result. They work over a range of 12–25m indoors. It is the almost universal system for exchange of computer data – and hence today's video signals. WiFi is used by smart TVs, PCs, tablets, gaming consoles, smartphones, broadcast receiver boxes, Blu-ray decks, media players and others. The later the equipment the more likely it is to incorporate WiFi: all new Sky+ HD receivers, and almost all Blu-ray players, for instance, feature it.

In addition to the WiFi compatible set-top box/TV, all that's needed for WiFi communication is a wireless media streamer, hub or router. This is usually supplied by your internet service provider as part of their deal. This acts as the WAP (Wireless Access Point) through which many signals and services can be simultaneously routed.

A single router can deal with many WiFi communicating devices, but as their number increases the data speed may drop as the available bandwidth/spectrum is shared between them; this is rarely detrimental in

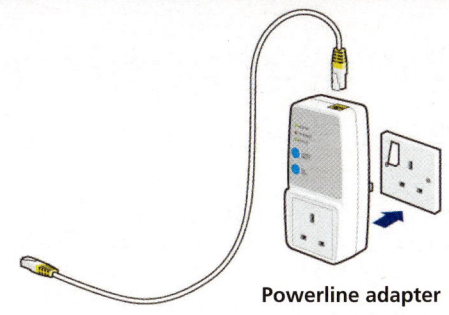

Powerline adapter

a home set-up, however, especially with the higher speeds now in use. WiFi Direct devices can communicate directly with each other without the need for a WAP. Setting-up is very simple: typically a network tab in the settings menu takes you easily through it, and an automatic scan of available networks finds and latches onto the router or other target device.

Interference and security
WiFi gear uses the same frequency band (2.4GHz) as several other devices, microwave cookers, cordless phones/headsets, baby monitors and fluorescent lamps amongst them, so there is a possibility of interference from these within your house and from your neighbours', generally taking the form of slow or no communication or intermittent data/picture drop-out. This can usually be overcome by resetting the operating channel from the installation menu. There is less crowding in the 5GHz band, for which later WiFi equipment is fitted, even having auto-band-select in some cases.

Security can be an issue, not generally with AV material, but with computer data and access to your home computer network. For this reason, some people prefer an Ethernet-based hard-wired network.

Ethernet
Ethernet cabling was originally designed as a carrier of data between computers. It can also carry digital television signals, and is frequently used nowadays to connect Smart TVs and set-top boxes to a home router. Cabling is cheap and the run length can be much more than with HDMI cables.

It's also useful where the distances are too great for WiFi, or the building fabric (solid walls or steel frames) restricts a WiFi signal.

Television Viewer's Guide

TV and audio troubleshooting

Save engineers' call-out charges with this simple DIY checklist.

These days TV sets are nothing more than elaborate home computers, purpose-designed for the job of receiving and processing digital sound and vision data, allied to a large screen or a projection system, and they can suffer similar problems to those affecting PCs.

Television set problems

With virtually any TV fault, from complete operational failure to a frozen picture, first try rebooting it by switching off, removing the mains power then plugging it back in again after five minutes.

In the case of a dead set (including no light-up, even of panel indicators) try switching on at the set itself (eliminates remote control faults) and then if necessary check the mains outlet socket and the fuse in the plug – it should be rated at 5A or more for a large screen model.

A bumping or screeching sound from within the set is a certain indication that it has developed an internal fault.

If the lights on the set come on but the picture and sound don't, switch between AV signal sources – aerial/tuner, disc player, game – and look for onscreen information. If they show, or if some services work, examine all your Scart, HDMI and other leads and their connections at both ends.

If broadcast reception has gone, confirm that the aerial or dish lead is present, correct and plugged in. If so try a retune or rescan, guided by the user manual. Go outside and check the aerial or dish itself, and the downlead, for damage and any obvious misalignment, see the next page.

Strange colours on screen can arise from a Scart or component connecting lead being loose or not fully pushed home. This cannot happen with HDMI connectors (they are generally all-or-nothing affairs!) but the loss of high-definition pictures, perhaps with a caption like 'out of range', can result from incompatibility of the digital drive signal: check the settings of the TV and the source box in the installation menu, described in their instruction manuals.

Set-top boxes

Digital receiver boxes, whether for Freeview or satellite reception, tend to be less reliable than TV sets. Thus it is more common for Freesat, Freeview and Sky boxes to need rebooting – again by disconnecting from the mains for a few minutes – to cure a wide range of symptoms, ranging from no sound or picture to some channels missing, strange foreign captions on screen etc. Check also, if necessary, the mains socket and plug fuse.

Some cases of malfunction after a running period of minutes or hours arise from overheating of the central processor in the box, generally caused by poor ventilation: see that the holes and slots in the case have free air circulation. A common symptom of overheating is cyclic freezing of the picture, even in the presence of a good input signal. For use inside a cabinet, cooling fans can be bought from, for instance, Satcure on www.satcure.co.uk

Viewers are often bemused to learn that a Sky+ box cannot play back the recordings already in its programme store unless the dish is present, correct and delivering a good signal and the subscription is up to date.

Disc players

As with the other equipment categories above, for a completely dead video deck, check the plug, fuse and mains socket, and for any malfunction at all try the effect of a mains power reset, described above.

Disc players are generally better behaved than TVs and set-top boxes; though they have moving parts they generally put in fewer working hours. A common fault in disc players is failure to read some or all discs, caused by a dirty or dusty scanning lens. It can often be cured by running a special cleaning disc.

Projectors

Most of the remarks made above for direct-view TVs are also true of projection models. However after a prolonged period of running, the lamp on a projector wears out, often

TV AND AUDIO TROUBLESHOOTING

heralded by onscreen announcements of impending life expiry or imminent failure. When the bulb finally dies there cannot of course be an onscreen indicator, but the message is given by body-mounted green/red indicators in some form. Instruction books will explains how to replace the lamp. Don't touch it, even cold or when new, with your bare fingers – you will damage it.

To prolong lamp life use if possible any low or 'eco' setting, and do not remove mains power for some time after switch-off.

Sound equipment

External speakers, soundbars and surround sound systems are usually quite trouble-free and reliable compared with vision display equipment. Even so they may occasionally need a mains reboot to cure problems.

During installation, relocation and possibly afterwards if any untoward sound effects, directional aberrations or strange intrusions are heard, it's necessary to set up and balance the surround decoder and loudspeaker-drive amplifiers. The simple procedure for this, which uses a microphone to evaluate the room acoustics and speaker placement, will be given in the user manual. If one speaker is silent or sounds bad, swap its feed with another to see whether the speaker or the amplifier is at fault.

Remote controls

If you're experiencing problems with a remote control the first thing to do is check its batteries. If that fails try rotating the batteries – this can rub off corrosion on the electrical contacts. If some buttons are not working the handset needs to be serviced or replaced.

Aerial and dishes

A loss of broadcast reception may be due to failure of the local transmitter, but this is less likely with a satellite. If your neighbour is still receiving the programmes you've lost, look up at your wall and roof. Is the satellite dish still there, undamaged and pointing in exactly the same direction as others close by? Is the download still attached to the LNB and intact all the way to its house-entry point? If so try your receiver on another dish then get yours checked by a professional.

Sky and other satellite signals are sometimes lost during periods of heavy rain.

Freeview retunes

With Freeview it's important to retune from time to time. If you are missing channels this is the first thing to do. If this doesn't work check that your aerial is level and looking in the same direction as those of your neighbours. Is the cable damaged? Is any signal booster/distributor plugged in and switched on? If the cable is damaged and in safe reach, you can repair it with F-type plug/socket connectors and a replacement section of co-ax cable, so long as you thoroughly wrap the joints in self-amalgamating tape.

Poor lip sync

Sometimes TV sound and picture can be out of synchronisation, caused by different time delays in the digital processing for each, and arising in the transmission path, the receiver or both. First try a re-boot, described above, and then see if the TV, set-top box or whatever has an audio delay preset in its installation or sound menu. Next if necessary check (with dealer or manufacturer) for a software or firmware update, applicable by various means. Try switching off picture – and sound, where applicable – enhancement features in the user menu, and rebuilding the planner in Sky receivers. Sometimes taking an audio feed from the TV or the HDMI switchbox to a surround amplifier or soundbar helps. With external audio equipment an adjustable sound delay box (e.g. Felston or Kramer make) can solve the problem.

General problems

If there is a frequent need to reboot one or more pieces of equipment to get them working properly the mains supply on site could be 'ragged', carrying frequent spikes or surges. Suppressors and surge-limiters, easily fitted, usually help.

Give TVs and mighty sound amplifiers their own mains wall sockets if possible, and feed smaller boxes from a good quality mains socket strip.

Whenever in doubt during fault diagnosis the best detection method is by substitution. Try a friend's or neighbour's equipment with your dish, aerial, TV or whatever, or take your suspect box to their house. Further advice for dealing with a range of interference and picture problems is given in the following pages.

Television Viewer's Guide

Interference and picture problems

This section describes some of the most common sound and picture problems and suggests possible causes and cures. Where the root cause can have several effects we have selected the most usual symptoms for inclusion here.

When a problem crops up there is often a simple solution. Try to isolate the cause yourself, or at least narrow down the range of possibilities. If the problem is with Freeview, retune your TV or set-top box. If the problem is with Sky, turn off the box, unplug it, then reconnect it after two minutes and restart.

If this fails check and inspect the leads, connections and aerial or dish; find out if your neighbour is suffering likewise; try another TV set or set-top box; see if the symptom crops up on all inputs/channels or services; and if necessary on Freeview confirm that your local transmitter is working properly at www.bbc.co.uk/reception/

To begin we shall look at the problems with standard-definition pictures and sound which arrive via the TV's Scart socket.

One colour missing
With an RGB vision link from a set-top box, DVD player etc. the three primary colour signals travel via separate conductors and connector pins. When one colour, red green or blue, goes missing it gives rise to an extreme hue error on the picture, which becomes cyan/turquoise, purple or yellow respectively, even if you turn the colour down to zero in the TV set-up menu. Check that the Scart plug is pushed fully home at each end, then if necessary try a different Scart lead. If wiggling the plug of a known-good lead makes the fault come and go there is probably a bad soldered joint behind the socket in the TV set or set-top box.

Vision crosstalk
It can happen that the picture is marred by shadowy movement and horizontal or vertical bars in its background, or an overlay of faint moving colours. These are the effects of crosstalk, in which vestiges of an interfering picture are superimposed on the wanted one. Again the Scart lead is most likely to be the culprit, being badly seated at either end, faulty or – in the case of a long one – not being of good enough quality for its purpose. An internal problem in the TV or source box can also be to blame here.

Left or right channel sound missing
The stereo sound channels pass into the TV in separate conductors and Scart pins, so as with the vision faults described above a badly-seated plug or faulty cable may be at the root of this; again bad soldering behind the Scart socket will generally be responsible if pressing and/or rocking the plug momentarily restores good sound. Check further by substituting the box or TV.

Audio crosstalk
The effect of this is the faint sound of an unwanted program in the background of the one being watched, mainly manifest when the volume is well advanced. Check the Scart cable and its connections as described above; cheap cables, especially long ones, can cause this symptom.

Sound distortion and vibration
Regardless of how the sound signal enters the TV, audio distortion, rattling and 'tizz' are common problems with modern TVs. Distortion can arise from too high a volume setting; if it happens at low sound levels the TV's speaker/s are probably damaged, usually as a result of setting the volume too high and thus overdriving the small or flat speakers. The best solution here is the use of a TV-

INTERFERENCE AND PICTURE PROBLEMS   83

external sound system as described elsewhere in this guide.

If certain notes and tones in the sound make an audible rattle or tizz make sure that nothing is touching the cabinet, then suspect that the rear cover or an internal component is resonant and vibrating. A dealer can help with this, or again an outboard speaker system hooked up.

Data corruption
Digital signals are rugged, and capable of delivering high quality pictures in the face of interference. If the TV signal quality deteriorates too far the picture usually either completely disappears or the decoding process breaks down and the picture degenerates into blocks like those shown here.

Another data-corruption effect is pictured above. This arose from impulse interference on a weak Freeview signal. It is spasmodic, and usually affects some channels more than others. These effects can arise from 4G interference, see below.

If the quality of the datastream is borderline there will be intermittent freezing and 'pixellation' of the image, with corresponding sound chirps and dropout.

There are various causes. With satellite systems, faults in the dish (including bad 'pointing'), the download or the receiver can be responsible. In exceptionally heavy rain

a satellite signal can fade to the point where the picture freezes. If it happens in light rain the signal may be marginal due to a small dish-pointing error. Freeview reception can be marred by impulse interference (from vehicles or electrical equipment) and strong signal reflections. Relocation of the aerial (e.g. further from the road) may help here.

'Local' digital feeds (from videotape equipment, DVD players etc.) can become corrupt when the heads, tape, disc or optical unit are dirty or misaligned. If the picture 'freezes' it may be cured by a mains reset. You do this by turning the power off for a couple of minutes and then back on again.

Lost signal
A complete loss of the broadcast signal results in a blank screen, usually with some explanatory caption like this one.

If you see this, it's unlikely that the satellite itself is responsible! Check that the dish hasn't been damaged, and that its line of sight is unobstructed; the LNB likewise. See that the cable is intact, and its plug is firmly fitted into the socket at the receiver. Ensure that any wall plugs are sound and tight. More serious faults can also cause a loss of signal. Cables deteriorate, especially if rainwater enters them. Set-top boxes can develop internal faults. Check the box by swapping it with one known to be working before calling in professional help.

Poor black level
Neither LCD nor plasma screens (especially early models) are good at achieving a true

Television Viewer's Guide

black in the darkest parts of the picture, though some makes and models are better than others, and the price paid is a major factor in this. This shot, taken from a mid-priced LCD screen, shows the effect of poor black level: the picture background should be the same as the border of the photo, which is totally black. A reduction in brightness (not contrast) level can help, while many LCD screens provide separate adjustment for ('electronic') brightness and backlight intensity, which latter has a major effect on perceived black level.

Recent LCD screens have LED backlighting, whose amenity to image-dependent dimming provides a great improvement in black rendering. In plasma screens the black level performance depends wholly on the design of the display panel.

Phosphor burn

Plasma screens use tiny phosphor dots to produce a picture, and the phosphor material is prone to burning, especially at high brightness and contrast levels. The result is a dull and subdued image, usually showing as a 'negative' of an icon or logo from a broadcaster or basic components of video games etc. On a large scale it appears when a full screen picture is displayed on a phosphor screen which has been under-scanned for a prolonged period.

Here the effect of this is graphically shown: the centre area of the picture has long been used for 4:3 display, and appears darker than the sides of the screen. There is no cure for this fault so it is important to take care while viewing, by ensuring that the screen is always full of picture; by avoiding bright stationary picture features lingering on it; and by setting low levels of brightness and contrast whenever there's any risk of screenburn.

Uneven panel illumination

Most easily seen with a uniform, near-black screen – between programs/titles; with no signal input and brightness slightly advanced; or from a pattern generator or disc – 'patchy' brightness can be, in mild form, a characteristic of the screen, especially among the lower-priced LED backlit types. If the effect is severe (generally in panels using the old-style florescent backlight) a fault is present, generally failure of a light-tube, its drive or connections.

LCD panel faults

In general thin screens are very reliable, and achieve good longevity. When they do fail various symptoms result: vertical or horizontal lines or bars, a negative image, a sparkling iridescence on isolated parts of the image, and others.

The picture above was shot from an LCD screen with an internal fault, and the screen was beyond repair, though the same symptom can arise from a defect in the (much less expensive) LVD printed-board assembly.

LCD screens are coated with a material which is rather soft and thus vulnerable to damage. Scuffs and scratches cannot be easily removed at home, though their effect may be reduced by carefully working in – with a microfibre cloth – Vaseline, or by rubbing with a pencil eraser.

Scuffs and scratches may be accompanied by mura defects, small uneven variations in the light intensity on screen. It comes from bruising or impact damage, and may be reduced by gentle surface rubbing with a microfibre cloth.

INTERFERENCE AND PICTURE PROBLEMS

The picture below shows another effect of a fault in the LCD screen layer or its immediate drive section, in which the number of brightness/colour levels has gone down.

Outside the guarantee it is generally cheaper, once the diagnosis is confirmed, to scrap and replace the whole set than to have an engineer replace the screen layer inside it.

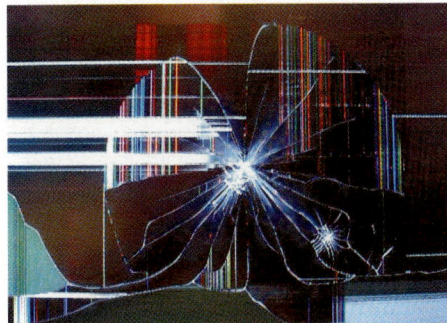

The picture above shows the effect of a broken LCD screen!

Software updates
All sorts of strange sound/vision symptoms, as well as performance problems, can be caused in digital receivers by software quirks. First try switching off at the mains for a few minutes. If that doesn't work it may well be that a software update is required. It can be installed by a tech-savvy user with the help of the dealer or manufacturer; over the air; over the phone line; or directly into cable tuners.

Internal faults
If your equipment develops what appears to be a serious internal problem with symptoms like buzzing, fizzing, arcing sounds or a strange smell from within the equipment, switch it off and unplug it at the mains before calling in a technician or taking it to a service

BBC INFORMATION

The BBC Reception Advice website has a section with information about reception problems.
http://www.bbc.co.uk/reception/
There is also a range of useful information about television and radio reception. This includes:
- Freeview, satellite and cable advice
- Retuning set-top boxes
- Freeview reception problems
- Aerial installation
- Satellite reception problems
- Useful links

PLANNED ENGINEERING WORKS

Digital UK is the organisation responsible for digital TV support in the UK. Its website lists planned engineering works.
http://www.digitaluk.co.uk/help_and_advice/engineering_works

OFCOM / BBC INVESTIGATION

There is information on Ofcom's website for those suffering from TV or radio interference or reception problems.
http://consumers.ofcom.org.uk/tell-us/
Please note that the BBC has taken over the responsibility of investigating interference problems. Check the link below to the Radio and Television Investigation Service first.
www.radioandtvhelp.co.uk
If you are not able to solve your problems you should talk to a local aerial installer to see if they can help. If problems persist you can ask a BBC specialist field engineer to investigate. The service is only available to domestic TV viewers and radio listeners and requires that you have a satisfactory external aerial. The service is not available if you live in a hotel, a block of flats or a housing estate with a shared aerial system.

Television Viewer's Guide

INTERFERENCE AND PICTURE PROBLEMS

centre. Continuing to run a TV, for instance, when it is obviously in distress can exacerbate the problem and incur a fire risk. You should talk to the original retailer where possible, or consult the manufacturer who can put you in touch with properly trained service agents.

There is one further type of interference that is not specific to one type of TV equipment: this is interference caused by 4G mobile phone services.

4G interference with Freeview

On Freeview TV transmissions in UHF channels 58, 59 and 60, and perhaps some others, there is a risk of interference, taking the form of pixilation, jerky or lost pictures, with loss or breakup of sound. In some areas retuning (see the TV or set-top box's instructions or installation menu) may be required – it's easy to do.

Where the problem occurs randomly and at short intervals the culprit will generally be a mobile phone in use nearby. Close to a 4G mast (say within a few hundred metres) the effect is more constant. Typically this affects the channels in the multiplexes nearest to channel 60. See the Freeview transmitter sites information in this guide, and the map at www.ukfreetv/maps/4g Local dealers and aerial riggers know from experience the areas and channels involved. The original estimates of viewer numbers affected were pessimistic, and only a few thousand have been disturbed, though the 4G rollout continues.

Locations

We have seen that interference is most likely near to a 4G mast or a mobile phone in use. The greatest risk areas are those with weak TV reception and a strong 4G signal. Many homes have an indoor, masthead or distribution amplifier dating from the analogue transmission era. These are vulnerable to 4G interference and they can be removed altogether in some areas now that the more efficient digital TV system is in use, and as home-wide video streaming and networking proliferates. In blocks of flats the distribution amplifier must necessarily stay and any 4G interference issue addressed by a professional technician – usually paid for by the management agent or resident's association.

4G Interference filter

Solutions

In most cases fitting a co-ax filter, pictured above, to the aerial downlead kills interference by blocking out the 4G signal. Free filters, for fitting by the householder or viewer, are available from at800, see below. It's important that the filter comes first in the chain, before any amplifier or distributor. This may be in a loft space, under the eaves or even at the top of an aerial mast – these need to be installed by a professional. In a very few cases filters cannot provide a cure; as a last resort it may be necessary to migrate to cable or satellite services to restore good reception. Free support and equipment for this can be had, see below.

Help and advice

A consortium called at800, financed and run primarily by the mobile network operators, provides free advice and on-site help. The latter limited to a single aerial-fed TV or receiver box in the home, including if necessary ladder work, performed by local specialist installers. For other situations, local dealers and riggers are ready to help, but on a chargeable basis. Filters, costing £5-£10, are freely available to buy.

Contact

at800 on Freephone 0808 1313 800 or www.at800.tv
The CAI (Confederation of Aerial Industries, www.cai.org.uk and 01923 803030/803203) can refer you to a local specialist in 4G interference.

4G future

4G, established in 2013, is spreading rapidly throughout the UK. One day it may take over the existing 600MHz TV band, calling for much retuning and in some cases new aerials and receiving equipment.

Archiving recordings made on video tape

If you have half-forgotten recordings on tape that you would like to preserve, don't delay transferring them to a digital format. Tape degrades over time and equipment such as VHS players and VHS/DVD combo recorders are no longer widely available.

Tape is not a good medium
for archiving. It has been superseded by better digital storage options. If you have VHS recordings that you wish to preserve don't delay transferring them.

How to store your tapes
If your tapes or videotape cassettes must be stored, see that they are properly labelled, that any record-safety tabs have been broken out, and that they are fully rewound. Videotape cassettes should be stored upright, flap facing into their boxes or sleeves (better, air-sealed in polythene food bags) in a cupboard or box at a normal and stable room temperature.

Before playing stored tapes, examine them for mildew, and acclimatise them to room temperature for two hours before play. Check the machine you are using for playback works with an unimportant tape before inserting a treasured recording!

The recording options
The two main options to consider are copying to DVD disc, or to a digital file that you can use, and even edit, on a personal computer.

The most popular method had been to copy to DVD. Blu-ray has been less popular as the discs are more expensive and home-based recorders not widely available.

How to copy onto DVD disc
The simplest and cheapest way to do it is with a stand-alone DVD recorder and your existing tape player/camcorder or VCR connected by a Scart lead and set to record and play respectively. Recording takes place in 'real time', so you will have to play the whole tape while the DVD is being recorded.

'Combo' VHS/DVD recorders
Widely available a few years ago, it's getting harder to find new equipment, though you can still find second hand and refurbished units for sale and on eBay.

If you do not wish to tackle the job yourself there are companies specializing in conversion. Expect to pay between £10-£15 to convert a basic 2.5 hours VHS tape. You will have to pay more if you want the disc edited.

Creating a digital file on a PC
If you have a PC you will need a video capture card in your computer to take the AV output from your VCR or camcorder. Alternatively an external device like Elgato's Video Capture, £70, will enable you to capture your tape's content on your computer digitally. Some camcorders have the capability to pass video signals through from other sources. This means you could connect a VCR to the camcorder's video-in jack, then output the same signal to your computer over Firewire or USB. Check the manual to see if this is possible with your camcorder.

Once you've captured the files digitally you can edit them in software such as iMovie or Windows Movie Maker. You can trim unwanted footage and add titles and graphics to the video footage. Once you've finished editing you can save the content in a format that you can view on your PC, laptop, tablet or smartphone.

Finally you can 'burn' one or more copies of your video from your computer to DVD to share with family or friends.

Another way to share content with family and friends, or even the wider world, is online via a free video sharing website such as YouTube or Vimeo. This can be a good way of sharing your videos without having to send them a DVD. In iMovie, for example, you simply choose Share, select YouTube, input your YouTube account details and select the size or quality of the movie you would like to upload.

Help with sight problems

The RNIB (Royal National Institute of the Blind) offers information and advice to over two million people with sight problems.

There is a wide range of sight problems, from slight loss of vision to complete blindness. For minor problems investing in a larger screen, changing the TV contrast, or simply darkening the room while watching may help. If the sight problem is more serious there are systems and products that may help.

TV Licence concessions
Children and adults who are registered blind can obtain a 50 per cent discount on the cost of the TV licence fee. You have to be registered blind and will need to supply your original local authority registration document. Refunds are available for previous years back to April 2000. Contact the TV Licensing Blind Concession Team on telephone number 0300 790 0366 for further information.

If you are aged 75 or over, you are entitled to a free TV Licence. The same applies if an elderly person lives with you.

Audio Description (AD)
Audio Description provides additional narration that describes all significant visual information such as body language, facial expression, scenery, action, costumes - anything that is important to conveying the plot of the story, event or image. It is available for some television programmes and is also provided in some cinemas and on some video discs.

Broadcasters (such as the BBC, Channel 4 and Sky) must add AD to 20 per cent of their programmes. AD is also available in Welsh on selected programmes. It is also available on Freeview, Freesat, Sky and cable. For further information contact the RNIB for a factsheet, or visit the RNIB website and search for 'audio description'.

Freeview
Freeview television delivers hundreds of hours of audio described programming every month. Audio Description is available on all Freeview HD products.

Freesat
Audio Description is available on Freesat. See www.freesat.co.uk for more information or you can contact Freesat on 0345 313 00 52.

Sky
Sky carries AD on 20% of its broadcasts. For quick and temporary AD press the help button on the Sky remote control. This menu will show whether subtitles and/or AD are available for the programme. Turn them on if available, and select save and exit. When you change channel these settings will be lost.

For permanent AD setting press the Services button on your remote. Then select Sky+Setup, followed by Subtitles. From here you can turn the Subtitles and Audio Description on and off. If you change Highlighted Programmes to show Audio Description, then you will see AD programmes highlighted when you search through the Sky EPG. Turn on Beep on Audio Description to hear a beep when you change to a channel showing an AD programme. Although some BBC programmes carry AD, they are not highlighted in the Sky planner.

Available on Sky services is a text-to-speech app for Sky+ receivers; it works with smartphones and tablets using the Android and iOS operating systems. More details at www.sky.com/accessibility

Virgin
Virgin Media carries AD on some of its programming. Check the Virgin Media website www.virginmedia.com for info.

Finding AD programmes
AD programmes are listed in Radio Times with 'AD' alongside their programme details. Sky have a weekly-updated list on their website http://accessibility.sky.com/ The 'tvhelp' website has useful lists of AD programmes carried on Sky, Freeview and Virgin Media, see www.tvhelp.org.uk

HELP WITH SIGHT PROBLEMS

BBC iPlayer
All on-demand content on desktop is subtitled and can be accessed from within the media player via the 'Subtitles' button.

Sign Language and AD (Audio Described) programmes can be found in the Sign Language and Audio Described pages accessed via the iPlayer Categories menu. You can also just simply search for your favourite programme and select the Sign Language or Audio Described version from the playback page.

For those on the move Audio Description is catered for by Android and iOS mobile devices: press the AD button below the programme information. The 'Watch with AD' button appears in magnified form.

Insight Radio
Insight Radio is RNIB's radio station. It broadcasts 24 hours a day on Freeview channel 730, also online at www.insightradio.co.uk In the Glasgow area it's available on FM radio 101.0 MHz. On an iPhone, iPad or iPod it comes via the free Insight Radio app, available from the App Store.

Equipment

TV remote controls
Some people with sight problems, especially those who are elderly or have reduced dexterity, have found the One for All range of remote controls easier to use than others: www.oneforall.com

Talking and smart TV Equipment
Panasonic has built 'Voice Guidance' as standard into some models of the Viera range. Voice Guidance announces onscreen information and important menus. Smart TVs from Samsung offer similar features.

Some internet TV set-top boxes also offer voice control. The Apple TV device works this way, utilizing its remote control and Apple's Siri voice control system. This lets you find content from iTunes and Netflix, but not all apps are supported – and it isn't really a mainstream TV device. Apple TV also has an iOS feature called 'VoiceOver' which is a screen reading technology and helps people with sight loss to find out what's happening on the screen.

Googles Nexus device offers similar voice searching using a microphone built in to its remote control.

Sky's specialised remote control
Sky has a remote control (Easy Grip), specifically designed to assist older customers, people with visual impairments and those with limited dexterity. It is now free. If you are ordering a new Sky installation it is worthwhile letting them know that you have sight problems as they may be able to offer additional help. Existing customers can call the Accessible Customer Service Team on 0344 2410 3333.

Talking Video remote control
Cobolt Systems sell a remote control, £60, that has been designed for people with sight problems. Cobolt Systems tel 01493 700 172 www.cobolt.co.uk

TV-based magnifiers
Several variable magnifiers are available for use with TVs having a Scart socket. Most are similar in appearance to a computer mouse and bring up on the screen, via a long cable, an enlarged image of the text, photo, label or whatever they are laid on. Prices go from £90 to £225. See www.rnib.org.uk > TV video magnifiers or search for on the internet.

Programme Information

Big Print - a weekly newspaper, published by the RNIB, which is printed with large bold type. It contains a 48 page TV Guide with programming schedules including BBC1, BBC2, ITV, Channel 4 and Five plus regional programming throughout the UK and Radio 2, 3, 4, and 5 and Classic FM programmes for the week. It costs £100 per year.

Braille version of Radio Times
Contact RNIB customer services on 0303 123 9999 for details of their braille, talking-book and emailed programme guides.

Further information - RNIB
105 Judd Street, London, WC1H 9NE tel 020 7388 1266 (shop and customer services). Helpline tel 0303 123 9999 www.rnib.org.uk

Television Viewer's Guide

Help with hearing problems

One in seven people have a degree of hearing loss, with an estimated eight million people each week using the subtitling service provided by UK TV broadcasters.

In this section we'll look at how to make the most of your television, stereo system or radio if you are deaf or hard of hearing. The Action on Hearing Loss (previously known as the RNID) information line is a good source of further information.

What system to choose depends on the severity of your hearing loss and whether you wear a hearing aid. We'll look at the options.

Loop systems
A loop system consists of a length of wire and a loop amplifier, which is attached to the wire. The wire itself is laid out at floor level around your room. The amplifier plugs into your television or other sound system either directly, for example via the Scart socket on your television, or by placing a microphone near the loudspeaker. The amplifier has a volume and tone control. You can buy and install a loop system for home use yourself.

How do you use a loop system?
If you wear a hearing aid, you need to switch it to the loop position, which is often marked with a 'T'. If you don't wear a hearing aid, you can still use a loop system but you will need a loop listener or a listening aid.

Infrared systems
An infrared system is an alternative to a loop system. It is easier to install because it doesn't need a wire loop.

The system consists of an infrared transmitter and an infrared receiver. The transmitter picks up sound from your television, stereo system or radio directly through a lead or using a microphone, and converts the sound into invisible infrared light. The receiver picks up the infrared light and converts it back into sound.

Two types of receivers are available – both have a volume control. If you have a hearing aid, you can use a receiver that has a neckloop attachment. The other type is used with headphones or stethoscopes. You have freedom of movement since both types of receiver are cordless and battery operated.

Listening aids
A listening aid amplifies the sound from your television or stereo system. It is sometimes called a sound amplifier.

It is a hand-held, battery operated device with volume control and perhaps tone control. You can plug it into the headphone or Scart socket of your TV set or stereo system.

You can then listen to the sound from your television or stereo system by plugging in a pair of headphones, earphones or a stetoclip into the headphone socket on the listening aid.

Headphones
Some televisions and sound systems have a headphone socket so that you can plug in headphones. The sound you hear through them may not be as loud as that provided by other equipment such as loop systems and listening aids.

Headphones are therefore suitable only for people with a mild hearing loss. You may need an extension lead to avoid having to sit too close to your TV set. You can buy headphones with volume control to adjust the sound level in each ear. Remember that using the headphone socket may cut off the sound for other people listening in the room with you. If you wear a hearing aid, it may be better for you to get a neckloop.

Wireless headphones
Wireless systems are a useful way to relay TV sound. Inexpensive systems are available from Maplin and others.

Using wireless headphones
One solution is to use a wireless sender or transmitter connected to the TV headphone socket. However this mutes the TV speakers, not ideal if others want to listen. Check the TV's instructions for any 'defeat' facility. Alternatively you can purchase a Scart audio

HELP WITH HEARING PROBLEMS | 91

sender from specialist dealer connevans.com tel 01737 247571, text 01737 644016. This plugs into a Scart socket and distributes the audio wirelessly, without muting the TV speakers, for use without disturbing others. Action on Hearing Loss offer a wide range of TV listening aids: see contact details below.

Using wireless speakers

Headphones can be uncomfortable to use for any length of time and can feel isolating. Instead you could use wireless speakers (available for £10-£20) to pick up the audio from the wireless sender. Speakers can be positioned near the viewer, or even mounted on the backrest of a chair.

How much subtitling is available?

The number of subtitles on programmes has increased dramatically in recent years. This has arisen from pressure from viewers around the country and from organisations. Now, over 90% of programmes on ITV1 and Channel 4 have subtitles, around 80% of those on Five, and all of BBC1 and BBC2.

Subtitles on Sky

There are two ways to set up subtitling on Sky. You can do it on a programme-by-programme basis, by pressing the help button on the remote control and following the onscreen menu. When you change channel the setting will be lost, and the subtitles will not display when you record on Sky+.

If you would like subtitles showing whenever they are available, or want to record them on Sky+, press the **Services** button, select **Sky+ Setup**, followed by **Subtitles**. From here you can turn Subtitles and Audio Description on and off. Don't forget to save the settings.

Subtitles on Freeview

Subtitles are carried on a range of Freeview broadcasts. There are many makes of Freeview set-top box - most support subtitles. Check your instruction manual for details.

Subtitles on Freesat

Subtitles are available for some programmes on Freesat.

Virgin Video Relay Service

Virgin Media offers its Video Relay Service to help hearing-impaired viewers communicate with its customer service team. To access VRS Virgin's customers need to go to www.virginmedia.com, click on the sign language interpreter icon and download a plug-in.

Subtitles on BBC iPlayer

You can get subtitles on programmes you stream or download. Downloads are available either via using BBC iPlayer Downloads on a laptop or desktop computer.

If subtitles are available, the button (S) will appear in the bottom right hand corner of the viewing screen once you've started playing the programme.

Android and Apple iOS mobile devices can subtitle iPlayer programmes. Users of Samsung Galaxy Note4 and S5 devices who have updated to Android may have difficulty accessing iPlayer. It's overcome by a firmware update: if necessary contact your mobile network provider.

Subtitles on disc

Many video discs have subtitles for deaf and hard of hearing people. Information about subtitles, including the language they are in, is on the disc's box. You have to select subtitles from an onscreen menu that appears when you play the disc. You do not need additional special equipment such as a closed caption decoder.

Signing on television

A limited number of programmes carry signing. Radio Times indicates any with an 'SL' alongside the programme information.

Buying Equipment

Action on Hearing Loss is the new name for the RNID. See its website for details of products, suppliers and costs. If you do not have access to the internet you can contact their information line. You will need to mention that you do not have access to the internet. Social services may help to pay for equipment. Contact your social worker for deaf people or your social services department directly.

Further information

Action on Hearing Loss. tel 0808 808 0123
Textphone 0808 808 9000
www.actiononhearingloss.org.uk

Buying a new aerial

Bill Wright explains what you should consider if you are in the market for a new television aerial.

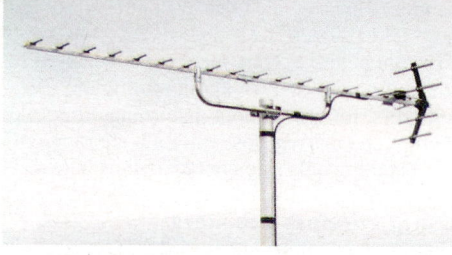

A good quality 18 element aerial

Whether you're musing about a DIY installation or you are definitely going to get a man in, it's useful to know something about that expensive bird perch that you're planning!

Firstly, will a new aerial actually be the best way to solve your TV reception problem? What about Freesat? See www.freesat.co.uk. Weigh the cost and the benefits of an aerial and all its paraphernalia against those of a Freesat receiver and dish. If you are in a very poor area for terrestrial reception Freesat could actually be cheaper, and will always give better reception, so if you have clear sky to the south-east the satellite option could be best for you. Don't be put off by any anti-Sky prejudice you might have: Freesat is nothing to do with the Murdoch Empire. Satellite is likely to be an easier installation job as long as you have a clear view of the sky to the south-east. The only real downside is that every TV set either needs built-in Freesat or a receiver box. A Freesat recorder will also serve as your receiver. Alternatively non-recording boxes are now very cheap.

Since we value all our readers I must also ask you DIY types to consider the continuance of your good health. Don't go climbing around at heights unless you are absolutely sure that you can do it safely.

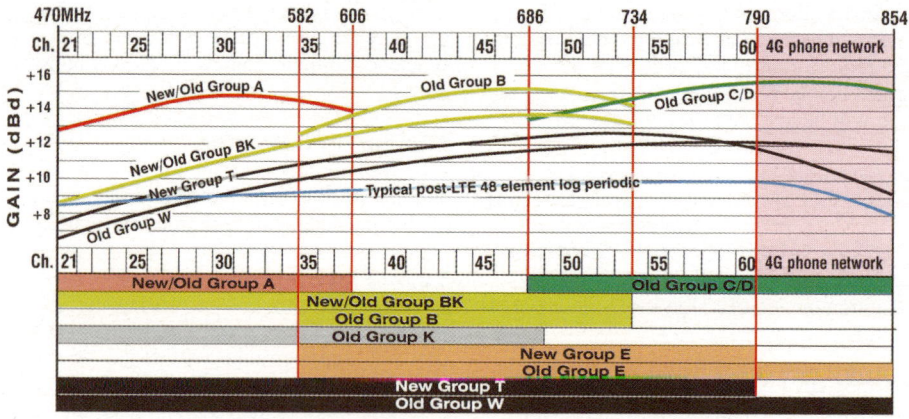

Channel groups and gain

The chart shows the channel groups currently in use, and the forward gain (sensitivity) of wideband and grouped versions of a typical good quality 18 element aerial. The gain figures are in dBd and are from our measurements, so are lower than most advertising claims. Some manufacturers quote dB (meaningless) or dBi, from which you should subtract 2.2dB (at least!). The channel groups each have a colour code, as shown on the aerial's end cap or elsewhere, and the colours are as shown here, but not all manufacturers follow this practice.

The narrower the band of channels the aerial is designed to cover the better its performance, especially in the lower part of the band. The colours shown are the channel group identification colour. The gain of a log-periodic aerial (see later) is also shown.

BUYING A NEW AERIAL   93

And even lofts can be dangerous, as can badly installed aerials (they fall on people). If in doubt, it's worth paying a professional installer. A good professional installer will know exactly which aerial you need once he's carried out rooftop tests, and he will be able to buy it and the accessories very cheaply, so the cost advantage of DIY might be much less than you think.

Not convinced? Still hell-bent on that aerial? Fair enough, I admire your spirit! Let's go for it then! Here's your crash course in aerials and installation!

Channel groups

Terrestrial TV signals are transmitted on channels 21 to 60 and aerials are available that cover the whole of the band. But because aerial performance decreases with the bandwidth covered there is an advantage to using an aerial that only covers the channels you actually need.

For this reason aerials are made for specific 'channel groups', as shown in the diagram opposite. If you're lucky the channels you need will fit into a channel group, so you can buy the aerial for that group. You will then benefit from its better sensitivity and directivity (the ability to reject interference). You can check the channels available from each transmitter by using list on page 113. The local relays generally fit everything neatly into one group, but many main stations don't. In particular the temporary HD multiplexes, COM7 and COM 8, will often force you to use a wideband aerial (Group T or W).

The actual channel groups in use are now a muddled mess, frankly. Originally we had Groups A, B, and C/D (and very rarely E, K and W). But channels 61 to 68 were taken away from TV broadcast and sold off to the phone companies. New groups T and E were introduced, but many manufacturers are still selling the 'old group' aerials, as well as the new ones, so both are shown on the chart opposite.

The disadvantages of using the old Group W, E, and C/D aerials are that the performance is optimised to include the 'dead' channels 61 to 68 at the expense of the lower parts of the band, and that if there happens to be a 4G phone mast nearby and in the same direction as the TV transmitter you are more likely to suffer interference from it. Some of the Group T aerials (especially log-periodics) have built-in filters to reduce pick-up of 4G ('LTE') signals.

The strongest case for using a grouped aerial comes when all the channels will fit inside Group A (channels 21–37). The sensitivity of a Group A aerial will far exceed that of a wideband when used for Group A signals, as you can see from the chart on page 92, and the directivity improvement is also impressive.

Try the main station

Now that analogue (with its ghosting etc) is no more, the main transmitting station could be the best bet even where there's a local relay. The main stations are the ones in bold in the list on page 113. Previously 'impossible' ones may now work perfectly. The relays only broadcast 'Freeview Lite', with fewer services than on the main stations.

Which channel group?

To find which channel group aerial to buy, decide on the transmitter you intend to point it at. Have a look at the aerials in your locality to see which way they point. Horizontal rods mean it's a main station; vertical mean it's a relay (almost always).

This guide, coupled with some vague idea of where you are in the UK, should be all you need to figure out the transmitter that covers your area (and thus the channels it uses), but if necessary further help can be gleaned from http://tx.mb21.co.uk/gallery/index.php (use the alphabetical index of sites) and www.megalithia.com/elect/terrain.html, where you can work out whether you have line-of-sight to the transmitter.

The transmitter ident given during tuning-in may give the name of the main transmitter for the region rather than the one you actually receive from.

Aerial type and quality

Cheap aerials are a waste of money. Unbranded ones with a perforated flat plate reflector should be avoided at all costs. The word 'digital' when it refers to aerials is meaningless by the way.

Many of the large complicated-looking aerials on the market are made to sell rather than to use! Look at your nearby rooftops. You will probably see a bewildering variety of aluminium artistry. Unless you are in a really

Television Viewer's Guide

BUYING A NEW AERIAL

Cheap, 'contract aerials', such as the one shown in the picture above, are best avoided!

A good quality 18 element aerial This is often the most cost-effective solution. This aerial has its elements aligned for a horizontally polarised signal

bad reception area (in which case you should be using Freesat) you do not need a huge aerial on your roof. If virtually every aerial in the neighbourhood is a massive thing on a tall pole, then fair enough, maybe you really do need something special. But, if only about a third or a half of the aerials on your street are monstrosities — and these will probably be on the poshest houses — the chances are that some of the local installers are super-salesmen, and reception isn't too bad at all. Most of the aerials you see date from when digital TV was low-powered, so nowadays a lesser installation might suffice.

Most aerials are of the 'Yagi' design, with a dipole, a row of elements called directors, and some form of reflector. The law of diminishing returns restricts the number of elements that can be usefully added to a Yagi. Elaborate designs with odd-shaped elements can only — by the laws of physics — work slightly better than those with straight rods. Once you go beyond a basic 18 element the performance improvement is slight compared to the increase in size, weight, and windage. If you really are struggling for every microvolt of signal then a massive aerial might be worthwhile, but don't expect a magical improvement over a standard 18 element array. And please don't image that a big aerial in the loft will equal a small one on the roof. It won't even come close.

The difference in cost between a ten and eighteen element aerial is so slight that you might as well always opt for the latter. If an 18 element aerial is appropriate for your

reception area you have a choice of good makes. Normally the channel group is a suffix on the aerial model name, for instance SG18A or SG18T.

Here are some of the good quality 18 element aerials available:
- Antiference TCX18 (Groups A, B, T)
www.antiference.co.uk/files/21.%20TCX.pdf
- Blake SR18
(Groups A, B, CD, E, K, T, W)
www.blake-uk.com/13-sr-yagi
- Triax digi 18 (Group T)
www.triax.uk/

Masts and brackets
A suitable support for an 18 element aerial is either a 6 foot long by 1¼ inch diameter aluminium mast, or if height is needed, a 12 foot long by 2 inch diameter aluminium mast. The tube wall should be no less than 16SWG (1.63mm). These specifications should be exceeded if the location is especially windy. In that case consider the use of an aluminium scaffold tube for a mast, as long as the brackets and masonry are really sturdy.

The 6 foot mast can be fixed to a substantial one-piece galvanised steel wall- (or chimney) bracket no less than 8 inches high and with arms made from angle rather than strip. If the mast has to stand away from the wall to clear an overhanging roof use a pair of large 'T & K' brackets. Don't use a cranked (bent) mast.

A 12 foot mast needs a pair of T & K wall brackets, or a double chimney fixing. The

BUYING A NEW AERIAL

brackets should be at least 2 foot apart and the masonry needs to be sound. A single chimney bracket with one lashing wire around the chimney is not adequate.

High gain aerials
High gain aerials vary from models similar in performance to an 18 element type up to really massive ones with perhaps 4 to 6dB more gain. Antiference, Fuba, Blake, Triax, Vision and others sell a wide range of high gain aerials, but not all channel groups are available from all manufacturers.

Log periodic aerials
Except in very weak reception areas one of the larger log periodic aerials can be a very good choice, especially if you need wideband reception. Logs are quite different to Yagis. They have lower gain but a very flat response right across the band. They are extremely directional and since rejection of interference matters as much as receiving the signal this is very important. Logs are small (about four feet long) and light, and do not catch the wind. A log plus a low gain masthead amplifier can be a very effective (and neat) combination. The manufacturers mentioned above all sell log-periodics. Avoid the very small logs though (those about a foot long).

High gain wideband aerials
'Wideband' and 'high-gain' are to some extent contradictions in terms, but sometimes a wideband aerial with the best possible gain is the only possible solution. All the different ranges of high gain aerials include wideband versions. Remember though, that the gain will be significantly less, especially on the Group A channels, than with a grouped aerial.

Log periodic aerial A vertically polarised log periodic TV aerial (top) shares a mast with a horizontally polarised FM radio aerial.

Where to buy?
I can guess where you're thinking of going on Saturday morning. Straight down to the DIY super-shed, am I right? Unfortunately these places only sell wideband aerials, and the quality and value can be very poor. Better to turn to a local specialist shop or Google for mail order suppliers such CPC UK or ATV Sheffield. Expect to pay £40 to £60 for everything you need for a good quality installation.

Surprisingly, many 'electrical wholesalers' are quite happy to sell to the public as long as the public knows what it wants and doesn't muck about asking daft questions and holding the queue up. So, if you know the aerial's model number and channel group a phone call followed by a visit can save you a lot of money. Have cash with you because some places don't take cards. Go at a quiet time, when all the tradesmen are hard at work. Just after lunch is good.

A few last words:
- Loft aerials are a severe compromise.
- Use CT100-type CAI approved cable always – it really makes a difference. The outer screen should be copper foil and copper braid, not silver paper.
- Good aerials are very directional and must be accurately aligned.
- UHF reception can be very different at locations only a few feet apart.
- Masthead amplifiers are good but are no substitute for an efficient aerial.
- Think hard about your safety and that of others before you even start.
- And if you really want to learn a lot more about aerials take a look at my website! www.wrightsaerials.tv/

DIY Disclaimer:
The advice given in this article is very general. We cannot anticipate details of any DIY work you might carry out. It is up to you to ensure that your work is safe. We can't accept responsibility for anything bad that might happen. If you have any doubts about your abilities get a man in!

Setting up a new aerial

Bill Wright guides you through the process of DIY aerial installation. This section will help you install or upgrade your television aerial. See also the information about choosing and buying an aerial on the previous pages. Note that Ofcom is planning to remove channels 49 to 59 from TV broadcasting by 2022.

Only consider aerial installation work if you are clear about the desired end, and are confident that you know what the work entails. In particular you must be able to work safely at height. Installation is a two person job. Do not risk injury or death to yourself and others by working without correctly used safety equipment. Before you start, have it clear in your mind exactly what you will be doing, and how you will ensure that every part of the work is safe. If you are at all unsure, use a professional installer. As well as having been trained to work safely, he should also know which type of aerial will suit your requirements.

Choosing an installer
The Confederation of Aerial Industries (CAI) can provide lists of its members, who also display the CAI logo in their adverts. However, a better way perhaps is to find a good installer in your area is by recommendation. Installers used long-term by health authorities, universities, local councils, etc, are a good bet.

Is it going to work?
First, make sure that reliable terrestrial TV reception is available in your area. The Freeview postcode checker will give you a good indication: www.freeview.co.uk/availability. The predictions are conservative so with a good aerial installation you might be lucky even if the postcode predictor says 'No!' Digital coverage is now at least as good as analogue ever was. But if you're in a really dodgy reception area, forget aerials: use Freesat.

There's no such thing as a digital aerial!
Don't assume that you need a new aerial because television is now digital. In most areas digital TV uses the same frequency band as the old analogue ones did, so older aerials will usually work as long as they are in good condition. In a few areas though you will need a new 'wideband' aerial. If you can receive some channels but not all this is probably the reason (but remember that the small local relays don't transmit every channel).

Improving poor reception from an existing aerial
Before you rush to replace an existing aerial there are a number of things you can check.
• Are you sure your set-top box isn't faulty? Is it tuned to the correct transmitter?
• Check the aerial connections inside the house, and in the loft if the cable is routed that way.
• Check for water in the aerial's junction box, corrosion, broken elements, damaged cable, etc.
• Align the aerial carefully.
• Remove disused splitters.
• Re-locate the aerial. Signal strength can vary a lot over quite a short distance. Can you re-locate the aerial so it isn't 'looking' through an obstruction such as a tree or a building?
• Increase the height of the aerial. Unless you can see the transmitter this is likely to increase the received signal strength.
• Use a masthead amplifier. Normally the lower gain, fully screened ones are best.
• Replace the cable with CAI benchmarked coax.

Choosing the aerial
A list of CAI tested and 'benchmarked' aerials is available from the CAI website. See also Bill Wright's article on page 92 for detailed information about your choice of aerial. Essentially you need to decide whether you need a big aerial or a small one, and whether to get a wideband one or a grouped one. The channels currently in use are numbered 21 to 60, although it is likely that channels 49-59 will disappear by 2022 and everything will get shuffled about yet again. So if you want

SETTING UP A NEW AERIAL

to be really future-proof you might decide to use a wideband aerial even if this is a performance compromise. But there again, will we need aerials at all in 2022?

The installation of a new aerial

The aerial should ideally be installed outside so it has clear line of sight to the transmitter. This might not be obvious because the transmitter might be too far away to actually see, but try to avoid having trees and buildings in the signal path. Loft aerials are a severe compromise. Aerials low down on the house wall are likely to be screened from the transmitter by other houses. For the normal sort of house, good locations are right at the top of the gable or on the chimney.

TV signals are transmitted with one of two polarisations. These are Horizontal (H), for which the rods on the aerial need to be, yes, horizontal, and Vertical (V) for which the rods need to be, well, can you guess? Main stations are usually H and relays are usually V. If in doubt look at the neighbouring aerials. You'll find this in the information we list for transmitters in the guide on page 113 onwards, under the Pol (Polarisation) column.

You can see this with the images on pages 94 and 95. The 18 element aerial has its elements aligned horizontally, and the log-periodic aerial on the following page has vertical elements.

When assembling the aerial make sure all the rods have the same polarisation. If some are at right angles you've done it wrong. How do you get on with Ikea shelf units?

The aerial should be carefully pointed towards the transmitter. The shortest rods should be nearest the transmitter. The aerial should not be pointed upwards unless the transmitter is close and on very high ground. It should never be pointed downwards. Bear in mind that aerials on surrounding houses might be incorrectly aligned.

Use good quality cable

Good quality cable is important, but there's a lot of really terrible co-ax on the market which will impair the signal. Use a CAI benchmarked cable such as Webro WF100 or Triax TX100. These have a double screen with braided copper wire and copper foil to shield the inside conductor.

Installing the downlead cable

If the cable can move in the wind it will eventually break inside, so it needs to be carefully fixed. If it rubs against masonry water can get in, then run down the cable and destroy your receiver! Kinking and squashing of the cable by heavy-handed staple gun use will affect reception. Drill holes 'uphill' so water doesn't run into the house. To join cable use two 'f' plugs and a line connector, then wrap with self-amalgamating tape if outdoors. Good quality outlet plates are fine but some cheap ones have a multipurpose circuit board that loses signal. Cheap flyleads are bad; make your own with good cable and 'f' plugs.

Masthead amplifiers

Masthead amps are powered from down below via the co-ax feeder cable, either from the accompanying power unit or from the TV set (it's a menu option). They are very effective where signals are weak, but are not a substitute for an inferior aerial. Use a low gain amplifier (9 to 16dB) unless the cable is over 30 metres, and even then use high gain amps with discretion. Set-back amplifiers are unlikely to improve reception. Use an amplifier that has 4G (LTE) filtering if you have a 4G phone mast nearby.

Loft installation

Loft aerials receive a much weaker signal and there's the risk of interference from the mains cables. However there are obvious advantages, so you can always buy the aerial and try it in the loft, then move it outside if necessary. Keep the loft aerial away from other objects, and find a location where it is 'looking' through tile rather than brick. If the roof is foil lined, forget it.

Can I get away with a set-top aerial?

For short-term use a set-top aerial can provide a solution, albeit an imperfect one. In a mediocre reception area it might only provide unreliable reception of just a few channels. Get a short extension cable so the set-top aerial can be placed high up or close to a window: that can make a real difference. The Antiference Silver Sensor is a good product. Amplified set-top aerials will provide little if any extra performance.

Television Viewer's Guide

TV reception in shared properties

Bill Wright looks at TV signal distribution in blocks of flats and other shared properties.

Because individual aerials and dishes can be impractical or unsightly, most multi-occupancy blocks in the UK have a communal TV distribution system. This provides TV reception for all. In essence an aerial and dish feed into a 'head-end' which is usually in the loft above the stairs. The head-end filters and amplifies the signals and sends them down separate cables to each living room wall plate.

Who pays?
In the case of privately owned flats the system is normally administered by the management agent or by a residents' association, with routine maintenance costs being paid from the ground rent fund and a special levy being imposed for any major upgrades. In the case of rented flats and residential homes the landlord usually covers all costs.

Problems can arise of course. Typically, a minority of residents may be very annoyed at some shortcoming of the system, yet the majority seem blithely unaware of the problem, so nothing is done. The unhappy few install their own aerials and dishes (often then getting hassle from the landlord or whoever) and lose interest in the communal system. This is one way systems can fall into an unrecoverable state of disrepair, after which rooftop anarchy prevails.

Dual input receivers
Nowadays most receivers for both terrestrial (aerial) and satellite reception can record programmes, and are known as PVRs (personal video recorders). This makes no difference at all to the terrestrial signal requirement, but it means that there has to be two independent satellite feeds, otherwise the PVR can't do everything it should. The majority of properties built before 2003 – and many built after that date – only provide one feed, despite the fact that the extra cost would be minimal if the work had been done as part of major building or renovation work. Once the building is finished and occupied though, conversion to twin feeds can be dauntingly expensive, especially in a building where visible cables are banned.

The installation industry has responded in a variety of ways. The first generation of devices on the market were 'Stackers', which squeezed two sets of signals down one cable by lifting the frequency of one set so it was above the other. Stackers did provide a solution at reasonable cost, but were often temperamental because the installed cable was not designed to carry such high frequencies. After Stackers a variety of proprietary (and mutually incompatible) solutions appeared. These used data signalling to allow the head end to send only the required multiplexes.

The most up-to-date systems use fibre optics, with fibre all the way to the dwelling, and this technology can provide a virtually unlimited number of feeds, allowing the use of PVRs in several rooms. If you are involved in a complete new system rebuild, fibre is worthy of consideration.

COMMON SYSTEM PROBLEMS
Unwanted transmissions Even though the aerial might be correctly aligned it could be receiving signals not just from the correct transmitter but from others as well. These unwanted signals can fool your receiver into tuning them in, resulting in unreliable reception and possibly incorrect regional programmes. Identify the correct transmitter using this book then carry out a manual retune, entering only the correct channel numbers (some receivers don't allow this though.) It is now regarded as good practice to fit filters at the head end to remove any unwanted signals, so if the problem persists ask that this be done. A more directional aerial might also be needed.

Incorrect filters Any channel filters fitted at the head-end will have needed adjustment or replacement at digital switchover. If this wasn't done the usual effect is to degrade or remove some programme services ('channels') whilst leaving others working

TV RECEPTION IN SHARED PROPERTIES

Fig 1 shows a typical TV distribution system based on a 'head-end', which will serve the whole building.

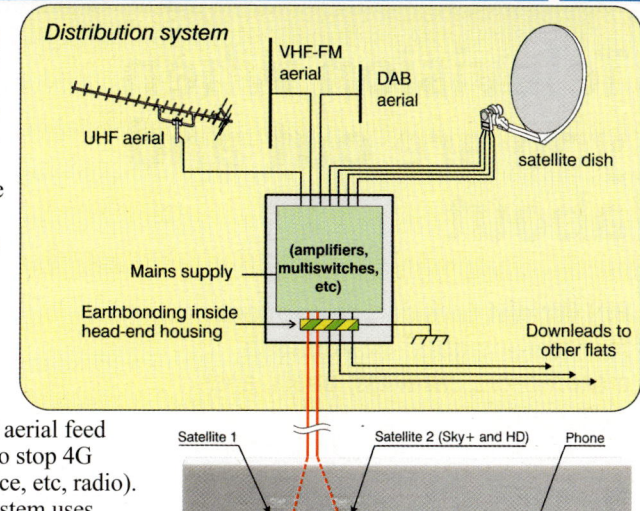

perfectly. A sure give-away here is when a set-top aerial gives better reception than the system on some but not all channels.

Interference If the system carries VHF FM radio or DAB radio it is essential that these frequency bands are rigorously filtered to remove out-of-band interference. Likewise the UHF aerial feed should have filters specifically to stop 4G phone signals and TETRA (police, etc, radio).

Wrong transmitter If the system uses a local relay only about half the Freeview services will be available. Ask if the system can be switched to a main transmitter.

Lack of head-end maintenance All systems needed adjustment in order to correct signal levels, following digital switchover. If this hasn't been done reception will most likely be unreliable. Faulty head-end amplifiers can cause strange problems that affect different receivers in different ways, which can be misleading.

Retention of obsolete equipment If the system used more than one TV aerial it is likely that this is no longer necessary following switchover. The old aerial and associated amplifiers should be disconnected because they might allow interference to enter the system.

Channel clashes The off-air channel line-up might clash with the in-house closed-circuit channels. This will wipe out some programme services and degrade the CCTV pictures. It is a simple matter to rearrange the channel line-up at the head-end.

Worn out systems If you suffered years of gradually-worsening reception before switchover, and now find post-switchover that things are not much better, then it's likely that the communal system has had its day. Typically, you would have had rather grainy analogue pictures and now you have unreliable digital ones. Pre-1985 systems will probably need serious work or even complete replacement. This isn't because digital TV is more demanding; it is because these systems

Fig 2 shows the features of an outlet plate that should be available in each flat.

have reached the end of their life anyway. TV cable, for instance, has a finite life, the length of which depends on its original quality and the dampness of the environment. Cables running in damp voids usually have a short life. These elderly systems can't carry satellite signals, and attempts to titivate them will prove futile. A complete new system is needed.

Poor reception in the locality Reception from the aerial might be unavoidably poor. Communal systems have advantages though: The building is likely to be tall, and with costs shared between many dwellings it is feasible to spend a good amount on the aerial. Even so it might be impossible to obtain really reliable aerial reception. The best solution might be to forget terrestrial reception and go for satellite, with residents advised to use Freesat or Sky. Even though they will need to buy receivers, satellite can be much cheaper and more effective than struggling on with hopelessly poor terrestrial signals.

Television Viewer's Guide

Television when you are out and about

Bill Wright considers the options for receiving good TV signals in motorhomes, caravans, boats, and even tents!

It used to be so simple. You arrived on site and put the aerial up. There were four channels at most and after a bit of aerial twiddling you accepted whatever reception you could get, and that was all there was to it, really. But now the options, and thus the decisions, have increased enormously. I'll try to guide you through the different ways of getting decent TV reception whilst you're out and about.

Use the laptop
Many travelling people regard a laptop as essential if only for backing up the photographs, and if you're one such you might not need to take a TV set at all. In a small van or boat the space saved can be a boon. A TV tuner dongle will convert your laptop into a TV set, although of course you will still need an aerial or a satellite dish. The dongle should be DVB-T2 capable.

The Internet
In a 3G or 4G coverage area the internet is a viable means of TV reception, using BBC iPlayer, ITV Player, etc. My own system is as follows. We have two smartphones, on different networks. Either can be used as a mobile WiFi hub to connect the laptop or the other phone to the Internet. This is very easy to set up; just have a look in 'Settings' for 'Portable (or Personal) Hotspot.' If the 3G/4G is weak try the phone near a window on each side of the van. In extremis you can put the phone into a plastic bag and reach through a vent to put it on the roof (but don't forget it when you drive off!) We share data usage between the two phones. My experience has been that for really rural locations EE provides by far the best 3G and 4G coverage in the UK, but all the networks have coverage maps on their websites.

WiFi
The larger campsites operated by the 'big two' clubs have on-site WiFi, as do many small sites, yacht harbours, and inland moorings. Prices vary a lot and in the UK especially they can be outrageous, so ask before you buy. Have a scout round the site for a WiFi base (usually a pole about 15ft high with a stubby little aerial at the top), and park or moor as close to one as possible.

To improve WiFi range and performance take a look at the clever gadgets on www.motorhomewifi.com. Pub campsites often have free WiFi inside the pub, and these devices can make it work in your van as well. Alternatively you can sit in (or near) the pub with your laptop and use iPlayer Downloads, then watch the programme later in the van or boat.

Aerials, dishes, and all that
There are several 'platforms', or over-the-air TV reception systems, so the first thing is to figure out which is best for you. The satellite options for the UK channels are Sky and Freesat. Terrestrial TV (TV from an aerial) is marketed as Freeview. If you opt for satellite you'll probably also have a minimal terrestrial set-up as a back-up.
So, consider your usage and circumstances:
• Do you have a mobile van or a static?

Television Viewer's Guide

TELEVISION ON THE MOVE

Obviously, mobile use needs an aerial or dish that can be deployed fairly quickly once on site. A static van or a boat on a near-permanent mooring can have a domestic-style semi-permanent installation.

- Where do you go? For very remote or mountainous areas of the UK, or anywhere abroad, you should plan to have satellite as your primary TV reception method, with a cheap and simple aerial as a back-up in case there isn't a clear view towards the satellite. To check terrestrial coverage go to http://tx.mb21.co.uk/mapsys/anatv/index.php and click on the relevant part of the UK.
- Do you go to the same site or mooring a lot? If so, the quality of terrestrial reception there and whether there's a clear view towards the satellite both need to be considered. Some sites in remote areas have a private 'self-help' TV transmitter or a wired TV distribution system, so don't assume the worst until you've actually investigated.
- How important is TV to you? If the telly is just for the news and the weather there's probably no point in putting a lot of effort into getting perfect reception.
- How much effort are you prepared to put into getting good reception at each stop? If you like to time your lay-by breaks to coincide with Coronation Street then you need a reception method that is well-nigh instant.
- Lastly you need to think realistically about just what is a physically feasible installation on your van or boat – or tent!

By now you're probably thinking, 'So I need either a dish or an aerial, or maybe both. I need Sky or Freesat, or Freeview.' Take a look at Which Digital Platform to Choose in this TV guide, then come back here for a quick look at the pros and cons, for mobile home use, of each platform.

Terrestrial television

Now that analogue TV is no more in the UK (and pretty well everywhere else) the broadcasters have been able to increase transmission powers – so digital reception can be surprisingly easy. Digital coverage in the UK is now at least as good as analogue ever was, and in many places significantly better. The small relay transmitters that fill in the gaps only broadcast three of the six digital multiplexes though, so if you like to have a wide choice of channels in your rural idyll consider Freesat. Having said that, the missing channels are designed for dull couch potatoes and shopaholics; not get-up-and-go camping and boating types like you.

A UK-purchased DTT (Digital Terrestrial Television) box or TV set should work in most countries, apart from the EPG (Electronic Programme Guide). In case of difficulty with UK kit, there's always the supermarché or supermercado, where local basic set-top boxes will be available, and cheap. These will be compatible with your UK TV set because the connection will be via a Scart or HDMI lead (All HDMI leads work just the same: buy the cheapest they have). Be sure to get a terrestrial box: in many parts of Europe the free-to-air satellite boxes are dominant in the shops. In France 'TNT' signifies a terrestrial digital receiver box.

At home or abroad buy only receivers or TV sets that are DVB-T2 capable. This might be advertised as 'high definition' but many standard-definition channels are also broadcast in the DVB-T2 standard. DVB-T1 is on its way out.

Sky

If you are a keen subscription television viewer you will most likely take your home Sky receiver camping with you to avoid the hassle of getting your Sky card mated with a different receiver. It's a nuisance lifting the receiver out of the house and installing it in the van, but if you have Sky+ you will have continuity of recordings across the period when you are away. Remember, though, that Sky+ won't work without a dish installed, even if you only intend to watch recorded programmes! Sky Multiscreen customers might have to explain the loss of the phone connection to Sky. Remember that Sky prohibits reception of their channels abroad, theoretically anyway, so don't mention it.

Freesat

If you don't like paying Sky, and furthermore would like to be able to watch your recorded programmes without putting the dish up, Freesat is the answer. Don't confuse Freesat with 'Freesat from Sky', by the way. Look for the Freesat logo when you buy a receiver.

Television Viewer's Guide

There are few small-screen TV sets available with built-in Freesat. A £50 Freesat receiver is the cheap answer, but really you should consider an HD (high-definition) Freesat recorder. These record and playback in HD, and in a few years time anything less is going to seem very old hat.

If you aren't interested in the news and the latest episode of the soaps, but prefer the more timeless sorts of programmes, you might well simply build up a viewing library while you're at home, to watch whilst you're away. If you think that might be a good idea you will need to either record programmes in the van whilst at home, move your receiver between house and van, or have two receivers of a type that will transfer programmes via a USB stick. Recordings have saved the day for us more than once when we've landed in some God-forsaken place (sorry, rural idyll) with constant heavy rain and no satellite, DTT, or internet!

Non-Freesat free-to-air satellite

Freesat receivers are very much based on the Freesat EPG. If you just want to watch the normal mainstream British channels, that's fine. You'll find them all neatly listed along with programme details and times, and it's very user-friendly. But if you intend to travel beyond the reach of UK satellite TV, or if you have an interest in foreign or specialist channels, you might like to sacrifice the convenience of the Freesat EPG and buy a standard free-to-air digital satellite receiver.

These are cheap: consider Fortec Star, Manhattan, and EchoStar. These boxes list every available free channel, with the Freesat ones all mixed in with everything else. You can compile a 'favourites list' though, giving quick access to everything that's available. Again an HD recorder is the future-proof solution. Alternatively, buy a Freesat box with simple access to the non-Freesat channels.

Very basic aerials

One of life's rules says that the more effort you put in the better will be the result, and this really does apply to aerials on mobile homes. The spectrum runs from set-top aerials (instant installation; low cost; very poor results except in good reception areas) up to large outside aerials (laborious installation; high cost; generally very good results).

If television is very low on your camping priorities, you can use a small set-top aerial and enough cable to reach to the van window. Although TV signals will enter a metal-bodied vehicle, reception is usually much better with the aerial near the window that faces the transmitter, as long as it isn't metalised glass or has a metalised sunscreen film. The best set-top aerials are the log-periodic ones, such as the Antiference Silver Sensor, £19, or the Telecam TCE2000 or TCE2001, at around £20. Aerials with built-in amplifiers can sometimes help a little, but they are not usually worth the extra cost. Very expensive elaborate-looking set-top aerials are a waste of space and money.

Directional or omni?

Moving rapidly on to better reception solutions, let's look at 'proper' outdoor aerials. You might assume that all an aerial has to do is receive signals, but a really effective one will be directional, receiving from one direction only and rejecting everything else. The drawback to this, for mobile use, is that the aerial can't be fixed rigidly on the roof and forgotten because it has to be pointed in the right direction at every stop. So the big decision is whether to go for the convenience of an 'omnidirectional' (non-directional) aerial, which is normally fixed to the roof, or whether to use a directional one, which normally needs a mast of some sort.

An 'omni' cannot discriminate between the signal and interference, so a directional aerial can work much better. But omnis are 'fit and forget' – a massive plus. There's no faffing around with the aerial because there's nothing to faff with. You simply turn the telly on, tune it in, and hope for the best. For those who move camp every day and feel that setting up even a simple directional aerial would be too much trouble, an omni on the roof could be the solution.

Many omnis come with an amplifier, or 'booster' built in, and these amps can provide a limited but possibly worthwhile improvement. If you have a caravan or boat in which the (inaccessible) aerial cable has been installed by the builder, you can bet your life it will be poor quality cable that loses a lot of signal, in which case an amplifier at the aerial (not behind the telly) can work wonders.

Directional aerials
Omnis will really only work in medium and strong reception areas. If you want to be able to get decent reception wherever the locals can, then you will have to use a directional aerial. The aerial also needs to be adjustable for polarisation, which is simply whether the rods need to be horizontal or vertical.

The simplest installations have telescopic masts of some sort that fit into permanent brackets on the van. Look at the Image 420 aerial and any of the Vision Plus masts, both from Grade UK. Also see Maxview's wide range of caravan aerials and masts.

The best type of aerial to fix on your mast is the log-periodic. These are compact, wideband, highly directional, and need no assembling. A log-periodic plus a masthead amplifier is a very efficient solution. The worst aerials are the complicated ones with rods sticking out in all directions. These look impressive but the performance is disappointing and they are impractical for mobile use. The need to repeatedly assemble and dismantle is a chore and usually ends up with something breaking.

Masthead amplifiers
Mast-mounted directional aerials used in weak signal areas will benefit significantly from the addition of a masthead amplifier. The gain figure should be low; no more than 16dB. The amp fits near the aerial and is powered via the cable from a small mains power unit near the TV set. The amp boosts the signal to compensate for the losses on the cable. Consider the Proception PROMHD11M, the Labgear LMA 113F, and the Antiference UXF1-15.

Adjustable roof-mounted aerials
As you will have gathered, mast-mounted aerials require you to put your cagoule on and brave the rain in order to adjust the aerial while 'er indoors shouts 'Better, better, back a bit!' The solution comes in the form of the permanently roof-mounted directional aerial that can be adjusted from inside the van or boat. The undoubted leader here is the Status 530 range, which allows you to raise and

Directional aerial - the **Status 550**, £140, allows full adjustment of direction and polarisation from inside the van.

rotate the aerial and alter the polarisation from inside the vehicle. Believe me this is so much better than going out in the wet!

Statics and normally-moored boats
Semi-permanent installations will owe more to domestic techniques (as discussed elsewhere in this guide) than mobile ones. Mainstream aerials, masts, and amplifiers are the thing. 12 foot masts are often the norm, with decent high gain aerials at the top. If reception is really poor at your static van an aluminium scaffold tube will cost about £35 and will get your aerial 20 feet up in the air. You can sleeve a 1.5 inch aluminium aerial mast into it and gain another 8 feet. The fixing needs to be pretty good though! If there's no convenient masonry you might have to use concrete or even guy wires. Use a grouped aerial if possible. See the information on buying a new aerial on page 92.

Aligning the aerial
I just love sitting in my motorhome watching the new arrivals trying to set their aerials up. It's better than watching the telly! So don't you provide amusement for that clever-clogs on the next pitch! Approach the problem scientifically. Find out in advance the directions of the likely transmitters, and look for a pitch that doesn't have an obstruction in that direction. Typical obstructions are barns (especially metal ones), the toilet block, huge American RVs, and dense clumps of trees with low branches. Pitches on high ground are good. If there's a hill in the signal direction get as far back from it as possible.

Don't take too much notice of other people's aerials – not everyone reads this book so many of the aerials will be installed incorrectly (I'm writing this on a campsite and I've just done a straw poll!) Aerials on nearby houses are a much better guide. Use a compass in conjunction with the transmitter information in this guide book. 'V' means 'vertical polarisation' – the aerial rods should be vertical. You can figure 'H' out yourself! Local relay stations usually use vertical polarisation. The short rods on your aerial should be nearest to the transmitter.

Having set the polarisation and rough direction, next optimise reception by fine aerial adjustment. You can't do it for digital reception by looking at the picture quality – digital doesn't work like that. A battery-powered in-line meter is the solution. These meters don't tune to an individual channel; they merely look at the whole band and give a general signal level indication, which is all you need.

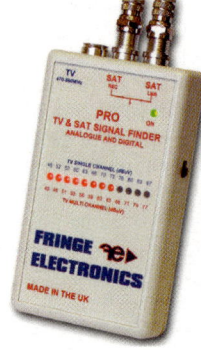

Fringe TV Signal Finder, £20, recommended for both satellite and terrestrial use.

Forget the cheaper meters and go for one of the Fringe TV Signal Finder range. They are well worth the extra money with significantly better detection circuitry and far higher meter resolution (12 LEDs instead of four).

If you seem to be getting signal peaks in two or more directions there will be several possible transmitters. Try them all if necessary, doing a 'factory reset' after each attempt to clear the receiver memory. If you have an adjustable 'booster', start with it at the minimum setting. All meters can be fooled by strong local 4G phone mast signals, especially if the 'booster' gain is high.

A domestic mini-dish on a tripod.

Satellite equipment and alignment

The dish can be mounted in a number of ways. A simple solution is a ground stand such as the tripod pictured above. Ground stands have two major advantages: they can be placed in a position where there isn't a tree in the way of the signal, and they are cheap. They take a few minutes to set up though, and they take up space in the van when travelling. Hint: a large plastic water container can be hung under the tripod to hold it down in windy weather.

Rooftop dishes vary from simple manual ones right up to horrendously expensive fully-automatic ones that find the satellite themselves. Maxview and Oyster are at the forefront of the mobile dish market. Remember that you will need a multi-output LNB for a receiver that needs two dish feeds.

Do your homework before you reach the site. Read David Sullivan's article overleaf in this guide. As soon as you arrive on site figure out which way is south east, and choose a pitch with no nearby trees in that direction.

After a bit of practice on site you'll be able to judge quite accurately which trees will obstruct the signal because you'll learn the elevation angle of the satellite. A small satellite meter and a magnetic compass are essential. Find the approximate direction from the compass, then use the meter for exact up-down and left-right alignment. Mark the up-down position on the bracket to make the job easier next time – it doesn't vary much

TELEVISION ON THE MOVE | 105

Maxview Winder - a simple manually-operated roof-mounted dish.

from place to place. If you get a high strength reading but no pictures you have the wrong satellite – move the dish round a tiny bit to the next one and try again. Some receivers will help with satellite identification. Finally, the first time you use the dish you should set 'polarisation offset' by rotating the LNB in its mount whilst checking the receiver's signal quality reading. For the Freesat/Sky satellite you should rotate the LNB about 30° anticlockwise as viewed when standing behind the dish looking towards the satellite. Once done this setting is permanent for the UK. From the LNB there should be a short cable, then a connection ('f' connectors) to the main downlead, so you can fit an in-line meter temporarily for alignment. All this sounds tricky, but it really isn't once you've learned how to do it.

If you opt for a ground stand keep a 30 metre CT100 cable extension in the locker so if you do end up under the trees you can put the dish in the clear.

Alignment of the manual rooftop dishes is much the same, except you don't get wet. I have one vital piece of advice though. If there isn't an ignition interlock, find an infallible way of reminding the driver to lower the dish before setting off!

Power

An inverter will produce mains voltage from a 12V or 24V battery, allowing you to choose from competitively-priced domestic mains-powered TV sets, receivers, and recorders. If you buy 12V products the choice is poor, and items tend to be over-priced. Only 'true sine' inverters are sure to be compatible with all your appliances. Take a look at TLC Electrical's range. A 300W unit (about £60) is adequate for a TV set and its ancillaries.

If all your TV equipment is mains-voltage powered, you can run it directly from site hook-ups when available. Organise your mains wiring so that the power source for the small appliances (the telly etc.) can be the inverter or the hook-up, at the turn of a switch.

Remember that 240V from even a tiny inverter can render you just as dead as 240V from the mains! An RCD is essential, and everything should be correctly installed. If in doubt get a man in!

Aerial and satellite cables

Use a CAI-benchmarked CT100-type cable (with a copper-coloured foil screen and copper braid) for every signal feed, both aerial and satellite. Foam-filled cable resists moisture better than air-spaced. If there are several TV sets a distribution amplifier should be used, with variable attenuation (signal level adjustment) on the input. Connectors should be 'f' types.

Where to buy

'Leisure' retailers tend to be expensive, so consider normal domestic aerial suppliers, manufacturer's own websites, DIY outlets, CPC, or aerial specialists such as ATV Sheffield. For installations on statics and normally-moored boats I would treat the caravan accessory market as the last resort. Ship's chandlers often have better quality caravan items than caravan shops!

Manufacturers and suppliers

Browse for the word 'aerial' plus any of the following: Status, Image, Maxview, Blake, CPC, Antiference, Fringe, ATV. Then look for 'leisure', 'camping', or 'marine'.

Television Viewer's Guide

How to set up a satellite dish

David Sullivan explains how to set up a Sky or Freesat satellite dish, useful in a caravan or when you are away from home.

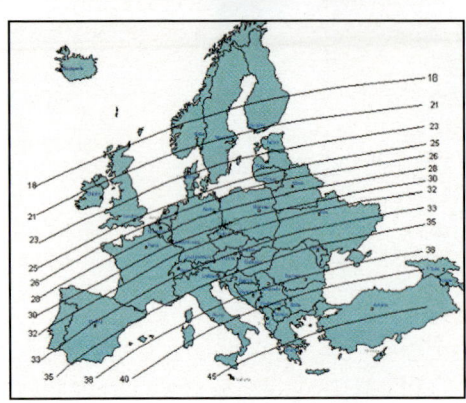

Approximate dish elevations in Europe

In previous pages, Bill Wright considers the options for watching TV on the move, whether through an aerial or via a satellite dish. I now want to take the satellite option a stage further and in more detail. Satellite reception, essential for watching in mainland Europe because the Freeview signals don't reach beyond the UK boundaries, requires a suitable dish and receiver, and the two most popular types of receiver are Sky and Freesat.

Sky and Freesat both work in similar ways. They display an EPG (Electronic Programme Guide, more usually called the TV Guide) with a full event schedule for the next seven days, and the viewer can select any programme to watch, or indeed to record.

Free-to-air satellite receivers

There is a third choice besides Sky and Freesat – generic free-to-air (FTA) receivers. These are becoming increasingly available in DIY stores and online and at ever-lower prices. They can be faster to set up but generally do not display a full EPG, being limited to a basic 'now and next' display. Many are dual-voltage, designed to work on 240V AC or 12V DC, which can be an advantage for caravans.

Sky and Freesat receivers

All current Sky and most Freesat receivers are mains only. There are a few 12V DC Freesat receivers but be careful - they are designed to run from a 240V AC supply via an adapter. Running them directly from a caravan 12V supply, where the voltage can vary considerably, can damage them and might invalidate the warranty.

The use of a Sky receiver does not require a subscription unless you want to watch any of Sky's own channels. All BBC, ITV, Channel 4 and Channel 5 services are free to watch (some HD services excepted), together with most news channels. If you opt for a new Sky subscription, an installer will come to your house and fit the dish for you, which for mobile use isn't really what you want.

Sky receivers can be purchased, without subscription and without installation, from independent suppliers, though not usually on the high street. There's also an active secondhand market in Sky receivers and you should be able to pick one up cheaply.

Freesat receivers are available from a number of manufacturers and are readily available off the shelf from high street suppliers. They are available in standard-definition (SD) and high-definition (HD), and as basic receivers or hard disk recorders. Be wary when buying a Freesat receiver because some unscrupulous online traders might try to sell you an FTA receiver instead. Look for the Freesat logo before parting with your money.

Whether you go for Sky, Freesat or FTA, the main job is to aim your dish in the right direction, and the technique is broadly the same in each case. In all cases, the signals carrying the programmes come from the same satellite system, so your dish will point in exactly the same direction.

Choice of dish

You also have a wide choice in dishes. The

HOW TO SET UP A SATELLITE DISH | 107

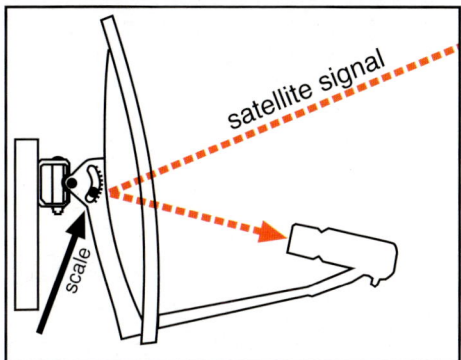

Dish Elevation. Dishes have a scale marked as an aid to setting the elevation.

black oval type you see on many houses up and down the country is cheap but has the disadvantage that it doesn't fold down for travelling. You'll pay around £60-£100 for a dish with a folding arm.

There are also types designed to be fixed on a caravan roof and some of these are fully automatic models that will lock on to the satellite without any manual intervention - expect to pay at least £1500 plus fitting! There is a cheaper alternative, the PAPSA (Portable Auto Positioning Satellite Antenna), available from a number of suppliers. It's free-standing so there is no fitting charge. Prices start at around £650. Avtex, the mobile TV specialist, markets a similar product under the Snipe brand name.

However, whichever type of dish you choose, there is a complication: the signals aimed at the UK can be more difficult to pick up in some parts of mainland Europe, requiring the use of larger dishes. A 60 cm dish will suffice throughout the UK, the Low Countries and most of France but will quickly prove inadequate as you travel further afield (see the section below headed 'The satellites have changed').

Tripods and signal meters

There are a few other items of equipment you'll need. A free-standing dish will need some kind of support, usually a tripod, and you'll also need a suitable length of satellite-quality cable. Finally, get yourself a signal meter. The basic Fringe meter shown on page 104 costs around £20. For more functionality search on Google for '3.5 satellite finder'.

There are many to choose from – the 3.5 refers to the screen size in inches – costing around £55. The biggest drawback is the instructions are generally poorly translated from Chinese to English. Sky+HD boxes, the only Sky boxes manufactured these days, are notoriously slow to react to the acquisition of a signal – so a meter is necessary.

It's also worth considering dishes equipped with an LED-light indicator which turns from red through amber to green when the dish detects a signal. It's called Easyfind and is incorporated into a number of makes of dish.

Aiming the dish

Aiming a dish isn't complicated but there is a trick to doing it, and even the use of a signal meter isn't foolproof. You won't get a signal by simply waving the dish around; the receiver has to process the signal before a picture or radio broadcast can appear for the first time, and this can take several seconds or even minutes depending on the make and type of the receiver. So if you just swing the dish around, you'll have already moved it out of the beam before the picture has a chance to appear. However, once you've got the hang of it, aiming a dish need take no more than a few minutes.

The dish has to be aimed at a precise point in the sky, with two coordinates involved – azimuth and elevation. For accurate alignment there is also a third, called skew.

Azimuth

Azimuth is the compass bearing to which the dish must be pointed. In the UK, it varies from about 140º to about 146º. In parts of Europe it can be as far east as 130º. There are a number of websites that will help you find the right direction – try dishpointer.com or sat-direction.com

Elevation

Elevation is the angle of the satellite above the horizon at sea level, but a complication is that most dishes are of an offset design. This means the signal has to strike the face of the dish at a downward angle so that it's reflected downwards on to the LNB (the small box on the end of the arm). Therefore the dish needs to be aimed below the elevation angle, not directly at the satellite, and I'll return to that in a moment. There are a few dishes, notably

Television Viewer's Guide

those manufactured by Multimo, that are designed to point directly at the satellite.

Skew

Have a look at most dishes on houses in Britain. You might notice that the LNB appears to be rotated clockwise a little. This is because of where the satellites are located in space. All satellites carrying TV signals are positioned above the equator, at such a height that their orbital speed exactly matches that of the earth. So they appear to hover which means that dishes don't have to move in order to track them. The satellite system that serves the UK is Astra 2, often referred to as Astra 28 because of its location at longitude 28.2 degrees east. Because it's not directly due south of us but further east and thus further round the curvature of the Earth, if we could see it, it would appear to be tilted. The LNB has to be rotated by an equivalent amount - this is called skew.

Remember what I said about setting the dish elevation low? For much of the UK, the dish face should point roughly straight towards the horizon (in northern Scotland the dish might even need to point down slightly towards the ground). The further south you travel, the higher you'll need to point the dish (roughly a degree for each 100 km).

Connecting the system

Now it's time to connect the system together. I would recommend you don't connect the dish without first disconnecting the power supply to the receiver because a small electric current travels up the cable to the dish and if you're careless in connecting the cable, you could cause a 'spike' sufficient to fry some of the electronic components!

Connect the dish to the receiver (via a signal meter if you have one) and switch on the power. Depending on the model of receiver, it might take a minute or two before it will respond to the remote control. Your next action will depend on which type of receiver you're using. With a Sky or Sky+ box press Services, then 4, then 6. A Sky HD box has a different menu layout and the sequence is also different. Press Services and then tab along the top command line to Setup, indicated by a spanner icon, and then to Signal on the line below. With a Freesat box it depends on the manufacturer but typically you'll press Menu, then Information or Setup.

If you have a free-to-air receiver, I'm afraid you're on your own, because the procedure will vary with each manufacturer - check the user manual. Basically you need to access the screen that allows you to select the system (often shown as Astra 28).

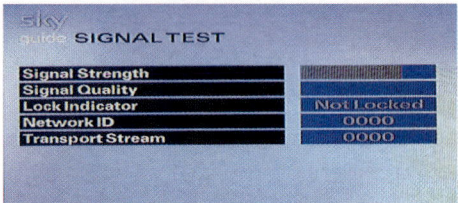

Figure 1A

A Sky box will display a Signal Test screen (Figure 1A above) with the Network ID and Transport Stream set to 0000 (zeros), or in the case of a Sky HD box, a list of random alphanumeric characters. A Freesat box will typically display an information screen (e.g. Figure 1B or 1C below) and the equivalent panels will be blank. Ignore any movement in the Signal Strength bar at this stage.

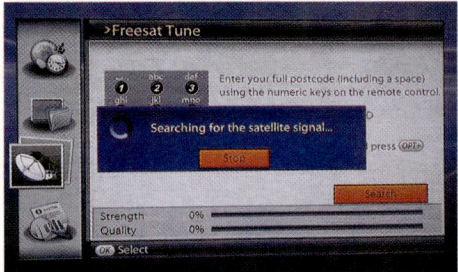

Figure 1B

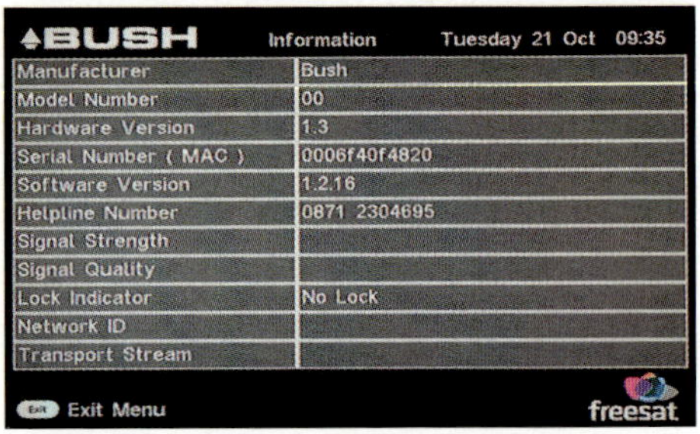

Figure 1C

HOW TO SET UP A SATELLITE DISH | 109

Astra 2 is roughly to the south east (the exact compass bearing will depend on where you are in Europe) so begin by turning your dish to the east. Now swing the dish slowly and a bit at a time towards the south. Sky HD receivers are very slow to react to the satellite signal so you will need to pause for considerably longer after each slight turn, and I would seriously suggest that the use of a signal meter is essential.

If you're using a meter, watch or listen for the indicator to rise which signifies that the dish is approaching the satellite beam. If the meter has a sensitivity adjustment, you might need to turn it up initially and then slowly turn it down as the signal gets stronger. Without a meter, turn the dish very slowly and watch the TV screen for the Network ID changing. On a Sky box, the value will change to 0002 (Figure 2A below).

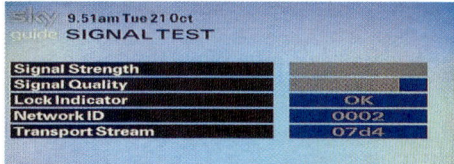

Figure 2A

On a Freesat box it might change to 003b (Figure 2B below). Some makes (eg Humax) don't show a Network ID at all but the signal bars will change from red through orange to green (Figure 2C below). On a FTA box, a picture will appear, normally behind the menu screen.

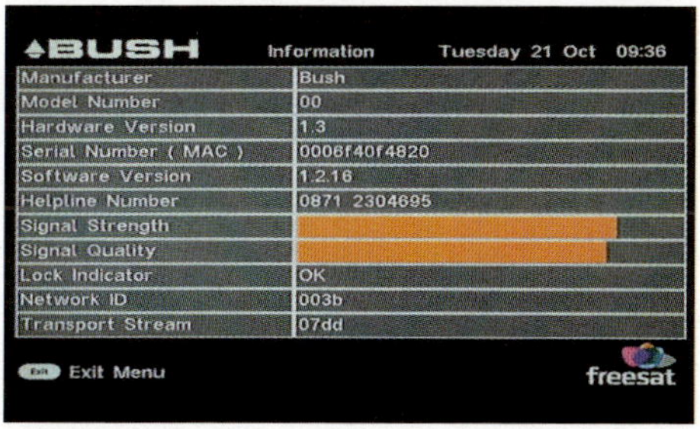

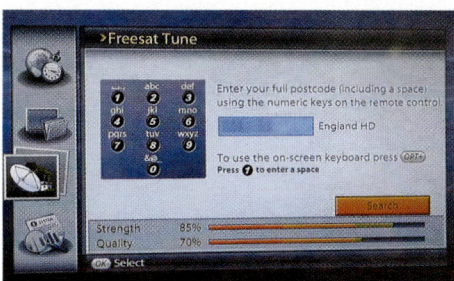

Figure 2B

Figure 2C

In the case of a Sky box, if you get a different Network ID value, you're pointing at the wrong satellite system. If nothing changes, try adjusting the dish elevation very slightly up or down and start again. Once you have the right values, you can then fine tune the dish alignment to get the highest possible Signal Quality reading. Anything higher than 40% should be OK, but above 50% will give some leeway for bad weather conditions.

The satellites have changed

Getting a signal in southern Europe is now difficult, at least as far as the main free entertainment channels are concerned. New satellites have been deployed to replace the ageing Astra 2 fleet. These new ones each have two beams – a narrow spot beam focused tightly over the British Isles and a pan-European one which covers much of mainland Europe. The spot beams carry BBC/ITV/Channel 4/Channel 5, and although the signals spread into the immediate mainland regions of northern France and the Benelux countries of Belgium, Netherlands and Luxembourg, they fade rapidly as you travel further north, east or south.

The German and Swiss borders are probably about the limit for a 60 cm dish, and to the south, the same dish will cope as far south as a line drawn roughly from Bordeaux to Lyon. Beyond those limits a larger dish will help but eventually the required size will become too large to carry.

If you're using a FTA receiver, it should work throughout most of Europe, although certain channels, including the four main UK broadcasters, will be missing once you get into southern or eastern Europe, and

Television Viewer's Guide

110 HOW TO SET UP A SATELLITE DISH

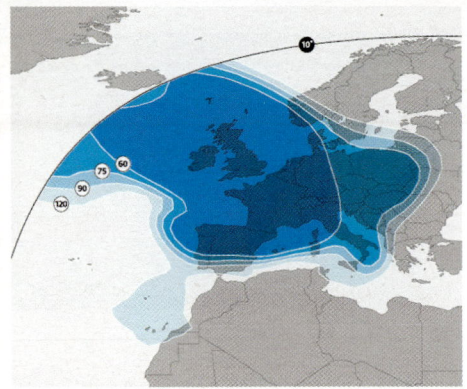

Astra 2 (EFG) wide beam

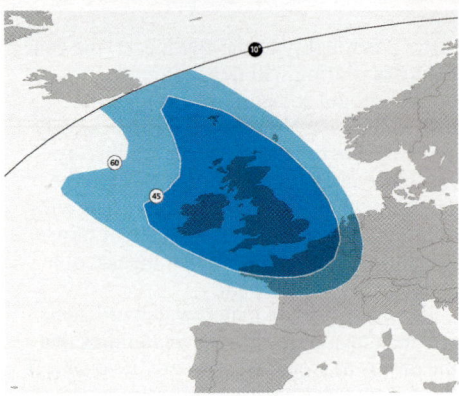

Astra 2 (EFG) narrow spot beam

obviously you won't pick up any subscription channels. Most news channels, except BBC News, will remain viewable.

A Sky digibox will work to a limited extent. You'll lose the same channels because they come from the same spot beams regardless of which type of receiver you use. However the Sky Electronic Programme Guide (EPG) which the receiver needs in order to work is carried on a European-wide beam, as are the subscription channels, so the EPG will be populated as normal and the subscription channels will remain available (see the coverage maps above).

A few free channels are also available to anyone with a Sky HD box and a subscription. Channel 5 HD and ITV2 HD, ITV3 HD and ITV4 HD are part of the basic Sky subscription package and are therefore available over the same coverage as the Sky channels. The SD versions of these channels are available on Freesat in the normal way but have restricted coverage across Europe.

The effect of the new satellites has now largely settled down but I will try to keep any changes up to date on my website.

If your receiver stops working

Most digiboxes are designed to be connected to a power supply at all times so that they can check for software upgrades. If you only use the box occasionally, you could miss some of them - a problem with Sky boxes if they are only used infrequently. If your receiver begins to behave erratically, or fails, it might need its software reloading. With a Freesat box, simply leave it in standby for a while and it should update itself automatically. Sky boxes don't, so you have to perform a manual download. Briefly (the full instructions are at www.satelliteforcaravans.co.uk/qanda.htm) the routine is as follows; power off at the wall. Press the Backup button on the fascia and, keeping it pressed while you switch the power back on. Keep the button pressed until all the fascia lights come on. The download takes a few minutes.

What are the legalities?

EU law dictates that no country can prevent you from picking up satellite signals but it doesn't require the broadcasters to make it easy. With Sky subscriptions, for example, you require the use of a Sky viewing card and officially you are not allowed to take that away from your home. Sky is unlikely to penalise you for temporary use of your card away from home, but remember you must not take a multiscreen viewing card with you.

With Sky's multiscreen (formerly multiroom) contract (whereby you get the same channel mix in two rooms for a few £s per month extra), you must keep the boxes and their viewing cards connected to the same phone line or WiFi connection permanently and Sky offers no relaxation of that rule. The company actively monitors these connections and transgressions may result in an increase of your direct debit to the level of two full-price subscriptions!

Further information and updates

www.satelliteforcaravans.co.uk

SKY'S INSTALLER SETUP MENU　111

Sky's Installer Set-up Menu

In the previous article David Sullivan mentioned performing a software update with a Sky set-top box. We would not usually recommend this as software updates happen automatically on Sky digiboxes as and when new versions are launched. This is generally every few months, to assist with ironing out bugs and adding new services. However, if you find that you are experiencing problems with your set-top box such as failed recordings or a loss of sound, or have not had your Sky box connected for some time a 'forced' software update might help.

In this section we will look at one of the other hidden features of Sky set-top boxes; Sky's Installer Setup Menu.

Make changes at your own risk!
There's a very good reason why this menu is hidden from general view - quite simply, if you incorrectly change some of the settings you can stop the set-top box from working, and wipe all your Sky+ recordings and Anytime content permanently.

Only make changes if you understand what you are doing, and make a note of the existing settings before changing anything.

Accessing the Installer Setup menu
On older Sky or Sky+ digiboxes, using the remote control you press:
the **Services** button, then numbers **4**, **0** and **1** - then press the **Select** button.

On a standard Sky HD or Sky+HD digiboxes: press the **Services** button, then numbers **0**, **0** and **1** - then press the **Select** button. Ignore changes on the TV screen and concentrate on pressing the buttons fairly quickly.

An image showing the Installer Setup menu is shown at the top of the page. Within it you will find the following sub-menus.

LNB Setup
This menu controls the LNB (Low Noise Block) which is the device on the end of the dish arm - it collects the satellite signal and passes it down the cable to the set-top box.

The Single feed mode sub-menu (only present on the Sky+HD box) can be changed if you want to set up your box where only one satellite feed cable is available. This can be the case with some communal dish installations or when caravanning.

Default Transponder
This enables you to change the default transponder settings for your digibox, useful, for example, if you are using your Sky box outside the UK in southern Europe.

Telephone Settings
Do not change this unless you need to, for example, when you have an exchange system and need to enter a dialling prefix to obtain an outside line.

RF Outlets
Here you can adjust the RF outlets or aerial out connections on the back of your Sky box.

Manual Tuning
Manual tuning allows you to add other channels to your Sky box which do not appear on the EPG.

New Installation
Don't play with this! It is used by installers to authorise your Sky card.

Rebuild Planner or Sky+
A planner rebuild helps to reclaim missing disk space where the planner is showing an incorrect amount of free space. It also reboots the box and this is probably its most common use as it can save having to root around at the back of your equipment for the mains switch.

Full System Reset
Use as a last resort as a full system reset permanently deletes all recordings and Anytime content and reboots the box.

Television Viewer's Guide

Freeview digital TV transmitters

Legend:
- ▲ Main Transmitter
- ● Transmitters also offering the 6 main multiplexes

Scotland & Islands:
- Eitshal
- Rumster Forest
- Rosemarkie
- Knock More
- Durris
- Angus
- Torosay
- Rosneath
- Craigkelly
- Black Hill
- Darvel
- Selkirk
- Keelylang Hill (Orkney Islands)
- Bressay (Shetland Islands)

Northern England:
- Chatton
- Fenham
- Caldbeck
- Pontop Pike
- Lancaster
- Bilsdale
- Olivers Mount
- Pendle Forest
- Keighley
- Idle
- Winter Hill
- Emley Moor
- Storeton
- Saddleworth
- Sheffield
- Chesterfield

Northern Ireland:
- Limavady
- Divis
- Brougher Mountain

Isle of Man / Wales:
- Douglas
- Llanddona
- Moel-y-Parc
- Long Mountain
- Blaenplwyf
- Presely
- Carmel
- Aberdare
- Pontypool
- Kilvey Hill

Midlands & Central England:
- Fenton
- Nottingham
- Belmont
- The Wrekin
- Sutton Coldfield
- Waltham
- Brierley Hill
- Bromsgrove
- Tacolneston
- Ridge Hill
- Malvern
- Lark Stoke
- Sandy Heath
- Sudbury

Southern England:
- Oxford
- Hemel Hempstead
- Crystal Palace
- Bluebell Hill
- Reigate
- Tunbridge W
- Dover
- Bristol KW
- Bristol IC
- Wenvoe
- Salisbury
- Hannington
- Guildford
- Heathfield
- Mendip
- Midhurst
- Hastings
- Huntshaw Cross
- Stockland Hill
- Rowridge
- Whitehawk Hill
- Caradon Hill
- Beacon Hill
- Plympton
- Redruth

Channel Islands:
- Fremont Point (Guernsey / Jersey)

FREEVIEW DIGITAL TRANSMITTER SITES

Freeview digital terrestrial TV (DTT) transmitters are listed below. Coverage has expanded, culminating with switchover completion in late 2012. The latest changes to Freeview involve the launch of new local TV services. There may also be some disruption and interference with the introduction of 4G mobile phone services.

Freeview multiplexes
At present six national multiplexes are broadcast from most transmitters. Three multiplexes are dedicated to Public Service Broadcasts (PSB) and three to Commercial (COM) services. In recent years two new multiplexes, **COM7** and **COM8** have launched. In addition local multiplexes have begun to appear, carrying the new tier of local TV channels.
Each multiplex can carry a range of TV or radio services. See the Freeview channel information on page 44 to find which multiplex carries a particular channel.

Freeview website
You can find a coverage predictor which checks your postcode to see whether you are likely to receive Freeview signals on the Freeview website www.freeview.co.uk/availability

Updates and latest transmitter news
Go to www.bbc.co.uk/reception/television and enter your postcode details. Your local TV and radio transmitters will be displayed, along with any reported problems.
Digital UK also provides a link to planned engineering work:
www.digitaluk.co.uk/help_and_advice/engineering_works

(Information updated by Mark Carver.)

NOTE - You will find the main transmitters, and some smaller ones highlighted in bold. These carry six or more Freeview multiplexes. Tune to one of these when possible, as they carry the most Freeview channels.

The powers shown are Effective Radiated Powers in kW. **POL** stands for polarisation - this shows the aerial group type and polarisation required for the transmitter. **V** - vertical and **H** - horizontal.

Multiplex names. **PSB** - Public Service Broadcasts, **COM** - Commercial

Transmitter	NGR	PSB1 BBC A		PSB2 D3&4		PSB3 BBC B		COM4 SDN		COM5 Arqiva A		COM6 Arqiva B		COM7 Arqiva C		COM8 Arqiva D		Pol
		Ch	kW	Ch	kW	Ch	kW	Ch	kW	Ch	kW	Ch	kW	Ch	kW	Ch	kW	
Aberdare	SO034013	24	0.1	21	0.1	27	0.1	25	0.1	22	0.1	28	0.1					AV
Angus	NO394407	60	20	53	20	57	20	54	10	58	10	49	10	31	3.3	37	5	WH
Auchtermuchty	NO214094	39	0.02	42	0.02	45	0.02											BV
Balmullo	NO426214	39	0.002	42	0.002	45	0.002											BV
Balnaguard	NN956511	39	0.002	45	0.002	42	0.002											BV
Blair Atholl	NN894658	43	0.008	46	0.008	50	0.008											BV
Carie	NN615572	24	0.02	27	0.02	21	0.02											AV
Crieff	NN814200	23	0.05	29	0.05	26	0.05											AV
Cupar	NO378139	41	0.004	47	0.004	44	0.004											BV
Dunkeld	NO046415	41	0.016	47	0.016	44	0.016											BV
Dunkeld Town	NO022430	23	0.002	29	0.002	26	0.002											AV
Grandtully	NN917527	49	0.002	54	0.002	58	0.002											CDV
Kenmore	NN774472	23	0.024	29	0.024	26	0.024											AV
Killin	NN602314	39	0.023	42	0.023	45	0.023											BHV
Lindores	NO251159	43	0.006	40	0.006	46	0.006											BV
Lochearnhead	NN594227	49	0.002	54	0.002	58	0.002											CDV
Menzieshill	NO368311	23	0.002	29	0.002	26	0.002											AV
Methven	NO016265	25	0.002	28	0.002	22	0.002											AV

Television Viewer's Guide

FREEVIEW DIGITAL TRANSMITTER SITES

Transmitter	NGR	PSB1 BBC A		PSB2 D3&4		PSB3 BBC B		COM4 SDN		COM5 Arqiva A		COM6 Arqiva B		COM7 Arqiva C		COM8 Arqiva D		Pol
		Ch	kW	Ch	kW	Ch	kW	Ch	kW	Ch	kW	Ch	kW	Ch	kW	Ch	kW	
Perth	NO108212	39	0.2	42	0.2	45	0.2											BV
Pitlochry	NN923565	25	0.025	28	0.025	22	0.025											AV
St. Fillans	NN663248	47	0.013	51	0.013	44	0.013											BV
Strathallan	NN860059	39	0.006	42	0.006	45	0.006											BV
Tay Bridge	NO430284	41	0.1	47	0.1	44	0.1											BV
Tummel Bridge	NN771601	39	0.02	42	0.02	45	0.02											BV
Beacon Hill	SX857619	60	20	53	20	57	20	42	10	45	10	51	10	33	7.1	34	1	WH
Ashburton	SX758687	24	0.002	27	0.002	21	0.002											AV
Bovey Tracey	SX818787	52	0.002	48	0.002	56	0.002											CDV
Brixham	SX921562	43	0.0036	40	0.0036	46	0.0036											BV
Buckfastleigh	SX742664	41	0.002	47	0.002	44	0.002											BV
Chudleigh	SX876789	41	0.002	47	0.002	44	0.002											BV
Clennon Valley	SX885596	52	0.002	48	0.002	56	0.002											CDV
Coombe	SX928736	24	0.002	27	0.002	21	0.002											AV
Dartmouth	SX875511	41	0.002	47	0.002	44	0.002											BV
Edginswell	SX886658	52	0.002	48	0.002	56	0.002											CDV
Halwell	SX781528	47	0.002	41	0.002	44	0.002											BV
Harbertonford	SX780559	52	0.002	48	0.002	56	0.002											CDH
Hele	SX912657	43	0.002	50	0.002	46	0.002											BH
Kingskerswell	SX873681	50	0.002	59	0.002	55	0.002											CDV
Liverton	SX811734	50	0.0025	59	0.0025	55	0.0025											CDV
Newton Abbot	SX851713	43	0.002	40	0.002	46	0.002											BHV
Occombe Valley	SX886625	24	0.02	27	0.02	21	0.02											AV
Sidmouth	SY136879	52	0.006	48	0.006	56	0.006											CDV
South Brent	SX690607	43	0.002	50	0.002	46	0.002											BHV
Tedburn St Mary	SX830941	45	0.004	49	0.004	42	0.004											BV
Teignmouth	SX936735	52	0.04	48	0.04	56	0.04											CDV
Torquay Town	SX915637	41	0.04	47	0.04	44	0.04											BV
Totnes	SX805594	24	0.002	27	0.002	21	0.002											AV
Belmont	TF218836	22	150	25	150	28	150	30	50	53	100	60	100	33	37	35	41	WH
Grimsby	TA280091	45	0.002	42	0.002	49	0.002											BV
Lincoln Central	SK984711	44	0.02	41	0.02	47	0.02											BV
Weaverthorpe	SE972716	55	0.009	59	0.009	62	0.009											CDV
Bilsdale	SE553962	26	100	29	100	23	100	43	50	46	50	40	50	31	18	37	18	WH
Aislaby	NZ863087	45	0.0076	39	0.0076	42	0.0076											BV
Bainbridge	SD935892	57	0.0076	60	0.0076	53	0.0076											CDV
Castleton	NZ693077	50	0.002	59	0.002	55	0.002											CDV
Eston Nab	NZ569182	52	0.003	51	0.003	48	0.003											BV
Grinton Lodge	SE048976	45	0.005	42	0.005	49	0.005											BV

Television Viewer's Guide

FREEVIEW DIGITAL TRANSMITTER SITES

Transmitter	NGR	PSB1 BBC A		PSB2 D3&4		PSB3 BBC B		COM4 SDN		COM5 Arqiva A		COM6 Arqiva B		COM7 Arqiva C		COM8 Arqiva D		Pol
		Ch	kW	Ch	kW	Ch	kW	Ch	kW	Ch	kW	Ch	kW	Ch	kW	Ch	kW	
Guisborough	NZ592168	57	0.01	60	0.01	53	0.01											CDV
Limber Hill	NZ789053	47	0.008	41	0.008	44	0.008											BV
Peterlee	NZ446410	45	0.002	42	0.002	39	0.002											BV
Ravenscar	NZ970012	56	0.033	58	0.033	53	0.033											CDV
Romaldkirk	NY974220	44	0.012	41	0.012	47	0.012											BV
Rookhope	NY926434	45	0.002	42	0.002	49	0.002											BV
Rosedale Abbey	SE730966	45	0.002	39	0.002	42	0.002											BV
Skinningrove	NZ715192	52	0.006	51	0.006	48	0.006											BV
West Burton	SE030880	45	0.002	39	0.002	42	0.002											BV
Whitby	NZ911095	55	0.1	59	0.1	50	0.1											CDV
Black Hill	**NS828647**	**46**	**100**	**43**	**100**	**40**	**100**	**41**	**100**	**44**	**100**	**47**	**100**	**32**	**43**	**35**	**39**	**WH**
Abington	NS938221	60	0.002	57	0.002	53	0.002											CDH
Biggar	NT016285	28	0.1	25	0.1	22	0.1											AV
Bridge of Allan	NS793985	26	0.005	23	0.005	29	0.005											AV
Broughton	NT129355	27	0.002	24	0.002	21	0.002											AV
Callander	NN670064	28	0.02	25	0.02	22	0.02											AV
Cathcart	NS566615	60	0.002	57	0.002	53	0.002											CDV
Clachan	NR766563	43	0.002	46	0.002	40	0.002											BV
Cumbernauld Village	NS754761	49	0.002	58	0.002	54	0.002											CDV
Deanston	NN712022	60	0.002	53	0.002	57	0.002											CDV
Dollar	NS951984	49	0.002	58	0.002	54	0.002											CDV
Dunoon	NS167771	27	0.002	24	0.002	21	0.002											AV
Easdale	NM754168	40	0.002	46	0.002	43	0.002											BV
Fintry	NS597889	27	0.004	24	0.004	21	0.004											AV
Gigha Island	NR643480	41	0.012	44	0.012	47	0.012											BV
Glasgow	NS565683	50	0.006	59	0.006	55	0.006											CDV
Glespin	NS821286	49	0.002	58	0.002	54	0.002											CDV
Haddington	NT538736	49	0.004	58	0.004	54	0.004											CDV
Kelvindale	NS555692	52	0.002	56	0.002	48	0.002											CDV
Killearn	NS483848	50	0.1	59	0.1	55	0.1											CDV
Kilmacolm	NS343691	27	0.006	24	0.006	21	0.006											AV
Kirkconnel	NS745149	49	0.05	58	0.05	54	0.05											CDV
Kirkfieldbank	NS862443	50	0.002	59	0.002	55	0.002											CDV
Largs	NS209594	45	0.005	42	0.005	39	0.005											BH
Leadhills	NS884149	49	0.002	58	0.002	54	0.002											CDV
Millport	NS166557	49	0.002	58	0.002	54	0.002											CDH
Netherton Braes	NS581575	27	0.002	24	0.002	21	0.002											AV
New Cumnock	NS612130	43	0.002	46	0.002	50	0.002											BV
Ravenscraig	NS252755	27	0.002	24	0.002	21	0.002											AHV
Rothesay	NS125690	28	0.4	25	0.4	22	0.4											AV
Rothesay Town	NS082648	50	0.002	59	0.002	55	0.002											CDV

FREEVIEW DIGITAL TRANSMITTER SITES

Transmitter	NGR	PSB1 BBC A		PSB2 D3&4		PSB3 BBC B		COM4 SDN		COM5 Arqiva A		COM6 Arqiva B		COM7 Arqiva C		COM8 Arqiva D		Pol
		Ch	kW	Ch	kW	Ch	kW	Ch	kW	Ch	kW	Ch	kW	Ch	kW	Ch	kW	
South Knapdale	NR837748	60	0.365	57	0.365	53	0.365											CDV
Strachur	NN094027	23	0.006	26	0.006	29	0.006											AV
Strathblane	NS555789	27	0.002	24	0.002	21	0.002											AV
Strathyre	NN559171	27	0.002	24	0.002	21	0.002											AV
Strathyre Link	NN581171	46	0.000	43	0.000	40	0.000											BV
Tarbert	NR858679	24	0.002	21	0.002	27	0.002											AV
Tignabeuaich	NR993743	45	0.018	49	0.018	42	0.018											BV
Tillicoultry	NS925971	60	0.002	57	0.002	53	0.002											CDV
Twechar	NS695754	28	0.002	25	0.002	22	0.002											AV
Uplawmoor	NS437563	49	0.005	58	0.005	54	0.005											CDV
West Kilbride	NS214483	44	0.2	41	0.2	47	0.2											BV
Blaenplwyf	SN569757	27	40	24	40	21	40	25	10	22	10	28	10					AH
Aberystwyth	SN587820	58	0.04	49	0.04	54	0.04											CDV
Afon Dyfi	SH844061	22	0.002	25	0.002	28	0.002											AV
Beddgelert	SH582476	50	0.002	59	0.002	55	0.002											CDV
Beddgelert Link	SH592490	27	0.002	24	0.002	21	0.002											AV
Bow Street	SN624845	44	0.017	41	0.017	47	0.017											BV
Corris	SH759067	45	0.002	49	0.002	42	0.002											BV
Cwrtnewydd	SN486476	44	0.004	41	0.004	47	0.004											BV
Dolybont	SN628889	57	0.02	60	0.02	53	0.02											CDV
Machynlleth	SH723003	57	0.004	60	0.004	53	0.004											CDV
Penryn-Coch	SN633844	50	0.014	59	0.014	55	0.014											CDV
Trefilan	SN562554	57	0.017	60	0.017	53	0.017											CDV
Ynys-Pennal	SN688983	44	0.008	41	0.008	47	0.008											BV
Bluebell Hill	TQ757614	46	20	43	20	40	20	45	20	39	20	54	20	32	4	34	5	WH
Chatham Town	TQ767675	57	0.005	58	0.005	52	0.005											CDV
Farleigh	TQ738530	56	0.0032	50	0.0032	53	0.0032											CDV
Bressay	HU503387	28	2	25	2	22	2	27	2	24	1	21	1					AV
Baltasound	HP635109	46	0.036	43	0.036	50	0.036											BV
Collafirth Hill	HU335835	45	0.8	42	0.8	39	0.8											BV
Fetlar	HU589914	47	0.25	44	0.25	41	0.25											BV
Fitful Head	HU347136	44	0.019	41	0.019	47	0.019											BV
Scalloway	HU398397	50	0.006	59	0.006	55	0.006											CDV
Swinister	HU440727	50	0.32	59	0.32	55	0.32											CDV
Voe	HU408634	57	0.002	60	0.002	53	0.002											CDV
Weisdale	HU379513	58	0.012	49	0.012	54	0.012											CDV
Brierley Hill	SO916856	60	2	57	2	53	2	50	2	59	2	55	2					CDV

FREEVIEW DIGITAL TRANSMITTER SITES | 117

Transmitter	NGR	PSB1 BBC A		PSB2 D3&4		PSB3 BBC B		COM4 SDN		COM5 Arqiva A		COM6 Arqiva B		COM7 Arqiva C		COM8 Arqiva D		Pol
		Ch	kW	Ch	kW	Ch	kW	Ch	kW	Ch	kW	Ch	kW	Ch	kW	Ch	kW	
Bristol (Ilchester Crescent)	ST577700	41	0.2	44	0.2	47	0.2	42	0.1	45	0.1	39	0.1					BV
Bristol (Kings Weston Hill)	ST547775	43	0.2	40	0.2	46	0.2	53	0.2	57	0.2	60	0.2					E/WV
Bromsgrove	SO948730	26	0.4	23	0.4	30	0.4	41	0.4	44	0.4	47	0.4					KV
Brougher Mtn	IH350527	28	20	22	20	25	20	21	2	24	2	27	2					AH
Brougher Mtn (RTE Multiplex)	IH350527	30	1															AH
Belcoo	IH090364	44	0.07	47	0.07	41	0.07											BV
Derrygonnelly	IH117514	44	0.004	41	0.004	47	0.004											BV
Ederny	IH238674	58	0.011	54	0.011	49	0.011											CDV
Lisbellaw	IH309410	49	0.004	54	0.004	58	0.004											CDV
Caldbeck: England	NY299425	25	100	28	100	30	100	23	100	26	100	29	100	32	32	35	36	AH
Ainstable	NY539466	49	0.02	45	0.02	42	0.02											BV
Bassenthwaite	NY206304	49	0.032	45	0.032	42	0.032											BV
Bleachgreen	NX984199	60	0.002	53	0.002	57	0.002											CDVH
Coniston	SD327966	27	0.016	24	0.016	21	0.016											AV
Crosby Ravensworth	NY619152	60	0.002	53	0.002	57	0.002											CDV
Crosthwaite	SD437900	48	0.004	52	0.004	56	0.004											CDV
Dentdale	SD727854	60	0.01	53	0.01	57	0.01											CDV
Eskdale Green	SD135997	25	0.002	28	0.002	22	0.002											AV
Glenridding	NY395172	57	0.002	53	0.002	60	0.002											CDV
Glenridding Link	NY386184	24	0.002	27	0.002	21	0.002											AV
Gosforth	NY069012	49	0.1	54	0.1	58	0.1											CDV
Grasmere	NY339056	55	0.004	50	0.004	59	0.004											CDV
Greystoke	NY450299	60	0.002	53	0.002	57	0.002											CDV
Haltwhistle	NY674627	59	0.4	50	0.4	55	0.4											CDV
Hawkshead	SD342959	26	0.012	23	0.012	29	0.012											AV
Kendal	SD540912	60	0.4	53	0.4	57	0.4											CDV
Kendal Fell	SD509930	46	0.003	40	0.003	43	0.003											BH
Keswick	NY278224	24	0.024	27	0.024	21	0.024											AV
Kirkby Stephen	NY777082	60	0.002	53	0.002	57	0.002											CDV
Lorton	NY155278	60	0.01	53	0.01	57	0.01											CDV
Lowther Valley	NY520199	46	0.005	50	0.005	43	0.005											BV
Millthrop	SD658786	52	0.003	48	0.003	56	0.003											CDV
Orton	NY618071	43	0.006	50	0.006	46	0.006											BV
Pooley Bridge	NY477234	46	0.003	50	0.003	43	0.003											BV
Ravenstonedale	NY733047	60	0.002	53	0.002	57	0.002											CDV
Sedbergh	NX967115	46	0.1	40	0.1	43	0.1											BV
St.Bees	SD607879	49	0.05	54	0.05	58	0.05											CDV
Threlkeld	NY313256	60	0.002	53	0.002	57	0.002											CDV

FREEVIEW DIGITAL TRANSMITTER SITES

Transmitter	NGR	PSB1 BBC A		PSB2 D3&4		PSB3 BBC B		COM4 SDN		COM5 Arqiva A		COM6 Arqiva B		COM7 Arqiva C		COM8 Arqiva D		Pol
		Ch	kW	Ch	kW	Ch	kW	Ch	kW	Ch	kW	Ch	kW	Ch	kW	Ch	kW	
Whitehaven	NX992123	43	0.4	40	0.4	46	0.4											BV
Windermere	SD383980	44	0.1	47	0.1	41	0.1											BV
Workington	NY001277	49	0.01	54	0.01	58	0.01											CDV
Caldbeck: Scotland	NY299425	27	100	24	100	22	100	23	100	26	100	29	100	32	32	35	36	AH
Ballantrae	NX089827	58	0.002	49	0.002	54	0.002											CDV
Barskeoch Hill	NX810616	55	0.4	59	0.4	50	0.4											CDV
Cambret Hill	NX524578	44	2.8	41	2.8	47	2.8											BH
Creetown	NX432559	59	0.006	50	0.006	55	0.006											CDV
Dumfries South	NX970741	43	0.02	46	0.02	50	0.02											BV
Glenluce	NX203569	57	0.003	60	0.003	53	0.003											CDV
Kirkcudbright	NX686506	21	0.002	24	0.002	27	0.002											AV
Langholm	NY358830	57	0.004	60	0.004	53	0.004											CDV
Minnigaff	NX406661	26	0.002	29	0.002	23	0.002											AV
Moffat	NT077050	45	0.002	42	0.002	49	0.002											BV
New Galloway	NX615788	26	0.02	23	0.02	29	0.02											AV
Pinwherry	NX183876	22	0.011	25	0.011	28	0.011											AV
Portpatrick	NX007545	58	0.06	49	0.06	54	0.06											CDV
Stranraer	NX111632	57	0.05	60	0.05	53	0.05											CDV
Thornhill	NX855891	57	0.1	60	0.1	53	0.1											CDV
Caradon Hill	SX273707	28	100	25	100	22	100	21	50	24	50	27	50	31	12	37	2	AH
Aveton Gifford	SX694474	44	0.002	41	0.002	47	0.002											BV
Compton	SX495563	44	0.002	41	0.002	47	0.002											BV
Croyde	SS446395	44	0.002	41	0.002	47	0.002											BV
Downderry	SX313542	53	0.06	60	0.06	57	0.06											CDV
Fowey	SX125507	53	0.002	60	0.002	57	0.002											CDV
Gunnislake	SX439719	46	0.006	43	0.006	50	0.006											BV
Ivybridge	SX631538	52	0.05	40	0.05	48	0.05											BV
Kingsbridge	SX721431	46	0.034	43	0.034	50	0.034											BV
Looe	SX252534	46	0.02	43	0.02	40	0.02											BV
Lostwithiel	SX100588	46	0.002	43	0.002	40	0.002											BV
Marystow	SX437829	45	0.002	49	0.002	42	0.002											BV
Mevagissey	SX011445	46	0.01	43	0.01	40	0.01											BV
Modbury	SX660514	50	0.002	59	0.002	55	0.002											CDH
Newton Ferrers	SX545475	50	0.002	59	0.002	55	0.002											CDV
North Hessary Tor	SX578742	50	0.003	59	0.003	55	0.003											CDV
Okehampton	SX586968	45	0.016	49	0.016	42	0.016											BV
Penaligon Downs	SX026683	45	0.02	49	0.02	42	0.02											BV
Plymouth North Rd	SX476552	46	0.002	43	0.002	50	0.002											BVH
Plymouth Weston Mill	SX454574	42	0.002	45	0.002	49	0.002											BV
Polperro	SX205508	53	0.002	60	0.002	57	0.002											CDV

Television Viewer's Guide

FREEVIEW DIGITAL TRANSMITTER SITES

Transmitter	NGR	PSB1 BBC A		PSB2 D3&4		PSB3 BBC B		COM4 SDN		COM5 Arqiva A		COM6 Arqiva B		COM7 Arqiva C		COM8 Arqiva D		Pol
		Ch	kW	Ch	kW	Ch	kW	Ch	kW	Ch	kW	Ch	kW	Ch	kW	Ch	kW	
Port Isaac	SW998805	50	0.05	59	0.05	55	0.05											CDV
Salcombe	SX753398	41	0.003	44	0.003	47	0.003											BV
Slapton	SX816416	60	0.25	52	0.25	48	0.25											CDV
Southway	SX478599	50	0.002	59	0.002	55	0.002											CDV
St.Neot	SX183676	42	0.002	49	0.002	45	0.002											BV
Tavistock	SX485716	53	0.02	60	0.02	57	0.02											CDV
Widecombe in Moor	SX725754	52	0.002	56	0.002	48	0.002											BH
Carmel	SN576153	60	20	53	20	57	20	54	10	58	10	49	10					CDH
Abercraf	SN851123	25	0.025	28	0.025	22	0.025											AV
Brechfa	SN504289	27	0.004	24	0.004	21	0.004											AV
Bronwydd Arms	SN414237	27	0.002	24	0.002	21	0.002											AV
Builth Wells	SO036528	25	0.005	22	0.005	28	0.005											AV
Cilycwm	SN777406	27	0.002	24	0.002	21	0.002											AV
Cwm Twrch	SN760106	24	0.003	21	0.003	27	0.003											AV
Cwmgors	SN705123	24	0.005	27	0.005	21	0.005											AV
Erwood	SO089428	60	0.002	53	0.002	57	0.002											CDV
Greenhill	SM924015	24	0.03	27	0.03	21	0.03											AV
Llandrindod Wells	SO018635	39	0.4	42	0.4	45	0.4											BV
Llanelli	SN510023	39	0.02	45	0.02	42	0.02											BV
Llansawel	SN618368	28	0.002	25	0.002	22	0.002											AV
Llanwrtyd Wells	SN899454	24	0.002	21	0.002	27	0.002											AV
Mynydd Emroch	SS775901	43	0.018	50	0.018	46	0.018											BV
Penderyn	SN957087	39	0.005	42	0.005	45	0.005											BV
Rhayader	SN985701	23	0.02	29	0.02	26	0.02											AV
Talley	SN639332	39	0.002	42	0.002	45	0.002											BV
Tenby	SS109994	56	0.02	48	0.02	52	0.02											CDV
Ystalyfera	SN779078	39	0.001	42	0.001	45	0.001											BV
Chatton	NU105264	45	20	42	20	39	20	44	10	41	10	47	10					BH
Berwick	NT980547	27	0.038	24	0.038	21	0.038											AV
Rothbury	NZ031997	55	0.01	59	0.01	50	0.01											CDV
Wooler	NT989276	28	0.002	25	0.002	22	0.002											AV
Chesterfield	SK383764	26	0.4	23	0.4	29	0.4	43	0.4	46	0.4	40	0.4					WV
Craigkelly	NT233872	27	20	24	20	21	20	42	10	45	10	39	10	33	11	34	11	WH
Aberfoyle	NS523991	55	0.018	50	0.018	59	0.018											CDV
Canongate	NT263736	55	0.003	50	0.003	59	0.003											CDVH
Grangemouth	NS921796	53	0.002	60	0.002	53	0.002											CDV
Kinross	NT097996	59	0.025	50	0.025	55	0.025											CDV
Newbattle	NT324651	50	0.002	59	0.002	55	0.002											CDV

Television Viewer's Guide

FREEVIEW DIGITAL TRANSMITTER SITES

Transmitter	NGR	PSB1 BBC A		PSB2 D3&4		PSB3 BBC B		COM4 SDN		COM5 Arqiva A		COM6 Arqiva B		COM7 Arqiva C		COM8 Arqiva D		Pol
		Ch	kW	Ch	kW	Ch	kW	Ch	kW	Ch	kW	Ch	kW	Ch	kW	Ch	kW	
Penicuik	NT252590	54	0.4	49	0.4	58	0.4											CDV
West Linton	NT164508	29	0.005	23	0.005	26	0.005											AV
Crystal Palace	TQ339712	23	200	26	200	30	200	25	200	22	200	28	200	33	43	35	40	AH
Alexandra Palace	TQ296900	49	0.07	58	0.07	54	0.07											CDV
Assendon	SU734856	59	0.002	55	0.002	50	0.002											CDV
Biggin Hill	TQ411587	39	0.002	45	0.002	42	0.002											BV
Biggin Hill Link	TQ417591	59	0.0004	55	0.0004	50	0.0004											CDH
Bishop's Stortford	TL499214	59	0.014	55	0.014	50	0.014											CDV
Cane Hill	TQ291588	58	0.005	49	0.005	54	0.005											CDV
Caterham	TQ343557	59	0.006	55	0.006	50	0.006											CDV
Chepping Wycombe	SU877911	44	0.02	41	0.02	47	0.02											BV
Chesham	SP956008	43	0.02	40	0.02	46	0.02											BV
Chingford	TQ380946	60	0.002	57	0.002	53	0.002											CDV
Croydon Old Town	TQ319647	52	0.007	56	0.007	48	0.007											CDV
Dorking	TQ169482	44	0.009	41	0.009	47	0.009											BHV
East Grinstead	TQ386360	43	0.023	55	0.023	50	0.023											WV
Edmonton	TQ345934	60	0.02	57	0.02	53	0.02											CDH
Farningham	TQ547660	52	0.013	48	0.013	56	0.013											CDV
Finchley	TQ251908	48	0.013	52	0.013	56	0.013											CDV
Forest Row	TQ438362	56	0.024	58	0.024	40	0.024											CDV
Gravesend	TQ656715	59	0.043	55	0.043	53	0.043											CDV
Great Missenden	SP905005	49	0.017	58	0.017	54	0.017											CDV
Greenwich	TQ408781	44	0.003	41	0.003	47	0.003											BV
Hammersmith	TQ232786	59	0.002	55	0.002	50	0.002											CDV
Hampstead Heath	TQ272854	47	0.002	44	0.002	41	0.002											BHV
Hemel HempsteadTown	TL054065	60	0.008	57	0.008	53	0.008											CDV
Henley	SU779822	58	0.02	49	0.02	54	0.02											CDV
Hertford	TL320137	49	0.5	58	0.5	54	0.5											CDV
High Wycombe	SU856942	59	0.1	55	0.1	50	0.1											CDV
Hughenden	SU856974	43	0.012	40	0.012	46	0.012											BV
Kenley	TQ329592	43	0.028	40	0.028	46	0.028											BV
Kensal Town	TQ245820	52	0.025	56	0.025	48	0.025											CDH
Lea Bridge	TQ373879	50	0.002	55	0.002	59	0.002											CDV
Marlow Bottom	SU841885	49	0.011	58	0.011	54	0.011											CDV
Micklefield	SU897933	57	0.002	60	0.002	53	0.002											CDV
Mickleham	TQ163538	58	0.018	49	0.018	54	0.018											CDV
New Addington	TQ378628	59	0.016	55	0.016	50	0.016											CDV
New Barnet	TQ265965	59	0.004	55	0.004	50	0.004											CDV
Old Coulsdon	TQ314587	44	0.002	41	0.002	47	0.002											BH
Orpington	TQ458652	59	0.003	55	0.003	50	0.003											CDV
Otford	TQ533603	60	0.006	57	0.006	53	0.006											CDV

FREEVIEW DIGITAL TRANSMITTER SITES

Transmitter	NGR	PSB1 BBC A Ch	kW	PSB2 D3&4 Ch	kW	PSB3 BBC B Ch	kW	COM4 SDN Ch	kW	COM5 Arqiva A Ch	kW	COM6 Arqiva B Ch	kW	COM7 Arqiva C Ch	kW	COM8 Arqiva D Ch	kW	Pol
Poplar	TQ382812	39	0.004	45	0.004	42	0.004											BV
Skirmett	SU777902	41	0.025	44	0.025	47	0.025											BV
St Albans	TL132069	57	0.022	53	0.022	60	0.022											CDV
Sutton	TQ255647	59	0.005	55	0.005	50	0.005											CDHV
Walthamstow North	TQ378897	49	0.002	45	0.002	42	0.002											BV
Welwyn	TL223161	43	0.03	40	0.03	46	0.03											BV
West Wycombe	SU839936	43	0.006	40	0.006	46	0.006											BV
Wonersh	TQ024454	44	0.021	41	0.021	47	0.021											BV
Wooburn	SU916873	48	0.02	56	0.02	52	0.02											CDV
Woolwich	TQ460793	60	0.126	57	0.126	53	0.126											CDV
Worlds End	TQ264773	46	0.005	43	0.005	40	0.005											BV
Darvel	NS557341	22	20	25	20	28	20	23	10	26	10	29	10	31	7	37	8	AH
Ardentinny	NS186864	44	0.014	41	0.014	47	0.014											BV
Arrochar	NN288046	27	0.002	24	0.002	21	0.002											AV
Ayr South	NS354187	45	0.002	39	0.002	42	0.002											BH
Blackwaterfoot	NR912291	43	0.008	46	0.008	50	0.008											BV
Bowmore	NR318590	47	0.013	41	0.013	44	0.013											BHV
Campbeltown	NR707192	60	0.025	57	0.025	53	0.025											CDV
Carradale	NR817376	44	0.006	41	0.006	47	0.006											BV
Catrine	NS529255	50	0.002	59	0.002	55	0.002											CDV
Claonaig	NR876565	50	0.015	59	0.015	55	0.015											CDV
Dunure	NS250150	46	0.002	43	0.002	40	0.002											BV
Girvan	NX211981	50	0.05	59	0.05	55	0.05											CDV
High Keil	NR680082	44	0.004	47	0.004	41	0.004											BV
Holmhead	NS566199	44	0.002	47	0.002	41	0.002											BV
Kirkmichael	NS354068	45	0.004	49	0.004	42	0.004											BV
Kirkoswald	NS231075	28	0.008	25	0.008	22	0.008											AHV
Lethanhill	NS438105	60	0.05	57	0.05	53	0.05											CDV
Lochgoilhead	NS194978	53	0.002	57	0.002	60	0.002											CDV
Lochgoilhead AD	NS190977	22	0.0002	25	0.0002	28	0.0002											AV
Lochwinnoch	NS337582	60	0.017	57	0.017	53	0.017											CDH
Millburn Muir	NS378796	42	0.05	45	0.05	39	0.05											BV
Muirkirk	NS710267	44	0.02	41	0.02	47	0.02											BV
Port Ellen	NR338452	25	0.03	22	0.03	28	0.03											AV
Portnahaven	NR179523	23	0.002	29	0.002	26	0.002											AV
Sorn	NS558260	46	0.002	43	0.002	40	0.002											BV
Troon	NS324315	49	0.004	58	0.004	54	0.004											CDV
Wanlockhead	NS874126	44	0.002	47	0.002	41	0.002											BV
Divis	IJ287750	27	100	21	100	24	100	23	50	26	50	29	50	33	12.5	34	12.5	AH
Armagh	IH891446	44	0.025	41	0.025	47	0.025											BV

Television Viewer's Guide

FREEVIEW DIGITAL TRANSMITTER SITES

Transmitter	NGR	PSB1 BBC A		PSB2 D3&4		PSB3 BBC B		COM4 SDN		COM5 Arqiva A		COM6 Arqiva B		COM7 Arqiva C		COM8 Arqiva D		Pol
		Ch	kW	Ch	kW	Ch	kW	Ch	kW	Ch	kW	Ch	kW	Ch	kW	Ch	kW	
Banbridge	IJ134460	40	0.002	43	0.002	46	0.002											BV
Bangor	IJ500817	55	0.002	50	0.002	59	0.002											CDV
Bellair	ID295153	56	0.008	48	0.008	52	0.008											CDV
Benagh	IJ261135	28	0.2	25	0.2	22	0.2											AV
Black Mountain	IJ278727	45	0.005	42	0.005	49	0.005											BV
Black Mountain	IJ278727	39	1															
Camlough	IJ055247	59	0.5	55	0.5	50	0.5											CDV
Carnmoney Hill	IJ336829	46	0.016	40	0.016	43	0.016											BV
Carnmoney Hill	IJ336829	48	0.016															BV
Conlig	IJ503783	45	0.013	42	0.013	49	0.013											BV
Cushendall	ID209275	46	0.005	50	0.005	43	0.005											BV
Cushendun	ID255349	27	0.02	21	0.02	24	0.02											AV
Draperstown	IH763955	45	0.002	42	0.002	49	0.002											BV
Dromore	IJ198535	59	0.002	55	0.002	50	0.002											CDV
Glenariff	ID235258	49	0.009	54	0.009	58	0.009											CDV
Glynn	ID401006	50	0.002	55	0.002	59	0.002											CDV
Kilkeel	IJ281180	45	0.4	42	0.4	39	0.4											BV
Killowen Mountain	IJ207174	27	0.005	21	0.005	24	0.005											AV
Larne	ID395037	45	0.4	42	0.4	39	0.4											BV
Leitrim	IJ253424	58	0.2	54	0.2	49	0.2											CDV
Moneymore	IH856827	46	0.008	43	0.008	40	0.008											BV
Newcastle	IJ362303	50	0.8	55	0.8	59	0.8											CDV
Newry North	IJ090284	43	0.002	46	0.002	40	0.002											BV
Newry South	IJ093255	44	0.004	41	0.004	47	0.004											BV
Newtownards	IJ501738	50	0.002	55	0.002	59	0.002											CDV
Rostrevor Forest	IJ189176	39	0.023	42	0.023	45	0.023											BV
Whitehead	IJ476930	52	0.002	51	0.002	56	0.002											CDV
Douglas (IoM)	SC373746	53	1	60	1	57	1											CDV
Beary Peark	SC295832	46	0.2	43	0.2	40	0.2											BV
Foxdale	SC292778	26	0.002	23	0.002	29	0.002											AV
Glenmaye	SC232803	60	0.01	57	0.01	53	0.01											CDV
Jurby	SC365991	46	0.04	43	0.04	40	0.04											BV
Kimmeragh	NX450000	44	0.1	41	0.1	47	0.1											BV
Laxey	SC437836	46	0.02	43	0.02	40	0.02											BV
PortStMary	SC206678	46	2	43	2	40	2											BV
Ramsey	SC453934	39	0.1	45	0.1	42	0.1											BH
Union Mills	SC343769	46	0.0025	43	0.0025	40	0.0025											BV
Dover	TR274397	50	80	51	80	53	80	55	40	59	40			48	40			CDH
Chartham	TR103561	21	0.05	24	0.05	27	0.05											AV
Dover Town	TR311430	26	0.02	23	0.02	30	0.02											AV

Television Viewer's Guide

FREEVIEW DIGITAL TRANSMITTER SITES | 123

Transmitter	NGR	PSB1 BBC A		PSB2 D3&4		PSB3 BBC B		COM4 SDN		COM5 Arqiva A		COM6 Arqiva B		COM7 Arqiva C		COM8 Arqiva D		Pol
		Ch	kW	Ch	kW	Ch	kW	Ch	kW	Ch	kW	Ch	kW	Ch	kW	Ch	kW	
Elham	TR188449	26	0.004	23	0.004	30	0.004											AV
Faversham	TR003602	21	0.02	24	0.02	27	0.02											AV
Folkestone	TR220359	26	0.1	23	0.1	30	0.1											AV
Horn Street	TR190359	44	0.02	41	0.02	47	0.02											BV
Hythe	TR169344	26	0.05	23	0.05	30	0.05											AV
Lydden	TR263458	43	0.004	40	0.004	46	0.004											BV
Lyminge	TR167428	22	0.005	25	0.005	28	0.005											AV
Margate	TR370701	50	0.2	51	0.2	53	0.2											BV
Newnham	TQ950571	21	0.01	24	0.01	27	0.01											AV
Ramsgate	TR385651	26	0.05	23	0.05	30	0.05											AV
Rye	TQ904198	54	0.05	46	0.05	56	0.05											BV
Turnpike Hill	TR153344	44	0.005	41	0.005	47	0.005											BV
Durris	NO763899	28	100	25	100	22	100	23	50	26	50	29	50	32	15	35	15	AH
Balgownie	NJ927104	46	0.008	43	0.008	50	0.008											BV
Banff	NJ687617	45	0.056	42	0.056	39	0.056											BV
Boddam	NK126415	45	0.002	42	0.002	39	0.002											BV
Braemar	NO108907	45	0.003	42	0.003	49	0.003											BV
Brechin	NO604590	46	0.002	43	0.002	50	0.002											BV
Ellon	NJ958311	45	0.002	39	0.002	42	0.002											BV
Gartly Moor	NJ547326	58	0.44	49	0.44	54	0.44											CDV
Gourdon	NO827709	50	0.002	59	0.002	55	0.002											CDV
Lumphanan	NJ587049	42	0.002	45	0.002	39	0.002											BV
Peterhead	NK112453	50	0.1	59	0.1	55	0.1											CDV
Rosehearty	NJ934663	44	0.4	41	0.4	47	0.4											BV
Tomintoul	NJ163209	46	0.002	43	0.002	50	0.002											BV
Tullich	NO379984	50	0.014	59	0.014	55	0.014											CDV
Eitshal (Lewis)	NB305303	26	20	23	20	29	20	25	10	22	10	28	10					AH
Ardintoul	NG832242	45	0.009	39	0.009	42	0.009											BV
Attadale	NG924392	28	0.002	25	0.002	22	0.002											AV
Badachro	NG779741	46	0.007	43	0.007	50	0.007											BV
Borve	NF648019	28	0.002	25	0.002	22	0.002											AV
Bruernish	NF719023	43	0.002	46	0.002	40	0.002											BV
Clettraval	NF751716	44	0.4	41	0.4	47	0.4											BV
Daliburgh	NF736216	60	0.006	57	0.006	53	0.006											CDV
Duncraig	NG827324	44	0.033	41	0.033	47	0.033											BV
Inverarish	NG557343	46	0.008	43	0.008	50	0.008											BV
Kilbride	NF752148	39	0.026	45	0.026	42	0.026											BV
Kinlochbervie	NC225560	46	0.016	43	0.016	50	0.016											BVH
Kylerhea	NG752206	44	0.01	41	0.01	47	0.01											BV
Lochinver	NC092222	46	0.002	43	0.002	50	0.002											BV

Television Viewer's Guide

FREEVIEW DIGITAL TRANSMITTER SITES

Transmitter	NGR	PSB1 BBC A		PSB2 D3&4		PSB3 BBC B		COM4 SDN		COM5 Arqiva A		COM6 Arqiva B		COM7 Arqiva C		COM8 Arqiva D		Pol
		Ch	kW	Ch	kW	Ch	kW	Ch	kW	Ch	kW	Ch	kW	Ch	kW	Ch	kW	
Lochmaddy	NF950727	28	0.04	25	0.04	22	0.04											AHV
Ness of Lewis	NB533603	44	0.006	41	0.006	47	0.006											BV
Penifiler	NG489417	45	0.008	49	0.008	42	0.008											BV
Poolewe	NG860820	44	0.003	47	0.003	41	0.003											BV
Scoval	NG180516	50	0.132	59	0.132	55	0.132											CDHV
Skriaig (Skye)	NG451408	27	0.32	24	0.32	21	0.32											AV
Staffin	NG495666	45	0.009	49	0.009	42	0.009											BV
Tarbert	NB154001	45	0.019	39	0.019	42	0.019											BV
Uig	NG383640	40	0.002	43	0.002	50	0.002											BV
Ullapool	NH142935	45	0.016	49	0.016	42	0.016											BV
Emley Moor	SE222128	47	174	44	174	41	174	51	174	52	174	48	174	32	55	34	51	WH
Addingham	SE076492	43	0.046	40	0.046	46	0.046											BV
Armitage Bridge	SE132133	49	0.002	58	0.002	54	0.002											CDV
Batley	SE239249	49	0.0025	58	0.0025	54	0.0025											CDV
Beecroft Hill	SE237350	59	0.2	55	0.2	50	0.2											CDV
Blackburn-in-Rotherham	SK389926	60	0.002	57	0.002	53	0.002											CDV
Bradford West	SE133345	55	0.0025	59	0.0025	50	0.0025											CDV
Brockwell	SK367707	45	0.002	42	0.002	49	0.002											BV
Calver Peak	SK232743	49	0.05	45	0.05	42	0.05											BV
Cleckheaton	SE184245	59	0.002	55	0.002	50	0.002											CDV
Conisbrough	SK516981	60	0.002	57	0.002	53	0.002											CDV
Cop Hill	SE058138	25	0.2	22	0.2	28	0.2											AV
Copley	SE080223	59	0.002	55	0.002	50	0.002											CDV
Cornholme	SD918264	49	0.008	58	0.008	54	0.008											CDV
Cowling	SD970432	42	0.0026	45	0.0026	39	0.0026											BV
Cragg Vale	SE003229	49	0.005	58	0.005	54	0.005											CDV
Cullingworth	SE075381	50	0.0026	55	0.0026	59	0.0026											CDV
Dronfield	SK362791	50	0.002	59	0.002	55	0.002											CDH
Edale	SK126843	60	0.002	57	0.002	53	0.002											CDV
Elland	SE126213	49	0.002	58	0.002	54	0.002											CDH
Grassington	SE015639	23	0.012	26	0.012	29	0.012											AV
Hagg Wood	SE148105	59	0.0066	55	0.0066	50	0.0066											CDV
Halifax	SE103242	24	0.1	21	0.1	27	0.1											AV
Hasland	SK406697	57	0.002	60	0.002	53	0.002											CDV
Headingley	SE299361	49	0.0024	58	0.0024	54	0.0024											CDH
Hebden Bridge	SD988267	25	0.05	22	0.05	28	0.05											AV
Heyshaw	SE170631	60	0.1	57	0.1	53	0.1											CDV
Holmfield	SE089295	59	0.0044	55	0.0044	50	0.0044											CDV
Holmfirth	SE140086	49	0.0052	58	0.0052	54	0.0052											CDV
Hope	SK170230	28	0.0024	22	0.0024	25	0.0024											AV
Keighley Town	SE065405	23	0.002	26	0.002	29	0.002											AV

FREEVIEW DIGITAL TRANSMITTER SITES | 125

Transmitter	NGR	PSB1 BBC A		PSB2 D3&4		PSB3 BBC B		COM4 SDN		COM5 Arqiva A		COM6 Arqiva B		COM7 Arqiva C		COM8 Arqiva D		Pol
		Ch	kW	Ch	kW	Ch	kW	Ch	kW	Ch	kW	Ch	kW	Ch	kW	Ch	kW	
Kettlewell	SD987680	42	0.026	49	0.026	45	0.026											BV
Longwood Edge	SE112161	59	0.008	55	0.008	50	0.008											CDH
Luddenden	SE048248	60	0.012	57	0.012	53	0.012											CDV
Lydgate	SD930250	23	0.002	26	0.002	29	0.002											AV
Millhouse Green	SE299028	49	0.002	58	0.002	54	0.002											CDV
Oughtibridge	SK307952	59	0.008	55	0.008	50	0.008											CDV
Oxenhope	SE028338	25	0.04	22	0.04	28	0.04											AV
Primrose Hill	SE143151	60	0.0056	57	0.0056	53	0.0056											CDV
Ripponden	SE043188	49	0.012	58	0.012	54	0.012											CDV
Shatton Edge	SK194814	58	0.2	54	0.2	49	0.2											CDV
Skipton	SD909517	39	2	45	2	42	2											BV
Skipton Town	SD998509	24	0.0026	27	0.0026	21	0.0026											AV
Stocksbridge	SK275991	49	0.0024	58	0.0024	54	0.0024											CDV
Sutton-in-Craven	SE004428	23	0.0024	26	0.0024	29	0.0024											AV
Tideswell Moor	SK149780	60	0.05	57	0.05	53	0.05											CDV
Todmorden	SD958241	39	0.1	45	0.1	42	0.1											BV
Totley Rise	SK322807	51	0.016	48	0.016	52	0.016											BV
Walsden	SD927235	60	0.01	57	0.01	53	0.01											CDV
Walsden South	SD937216	43	0.002	46	0.002	50	0.002											BV
Wharfedale	SE198485	25	0.4	22	0.4	28	0.4											AV
Wheatley	SE068264	49	0.0032	58	0.0032	54	0.0032											CDV
Wincobank	SK388919	59	0.002	55	0.002	50	0.002											CDV
Fenham	NZ216649	27	0.4	24	0.4	21	0.4	25	0.4	22	0.4	28	0.4	31	0.4	37	0.4	AV
Fenton	SJ903451	24	2	27	2	21	2	25	1	22	1	28	1	32	0.28	34	0.28	AV
Fremont Point	CI908280	44	3.2	41	3.2	47	3.2											BH
Alderney	CI865798	49	0.2	60	0.2	55	0.2											CDV
Gorey	CI985222	59	0.01	52	0.01	48	0.01											CDV
Les Touillets	CI581511	54	1.5	56	1.5	51	1.5											CDH
St.Brelades	CI852206	59	0.015	52	0.015	48	0.015											CDV
St.Helier	CI920195	59	0.04	52	0.04	48	0.04											CDV
St.Peter Port	CI610503	24	0.01	27	0.01	21	0.01											AV
Torteval	CI515482	59	0.02	50	0.02	48	0.02											CDV
Guildford	SU975487	43	2	46	2	40	2	48	2	52	2	49	2					BV
Hannington	SU527568	45	50	42	50	39	50	44	25	41	25	47	25	32	37	34	26	WH
Aldbourne	SU262752	27	0.002	24	0.002	21	0.002											AV
Alton	SU722386	57	0.002	60	0.002	53	0.002											CDV
Chisbury	SU274651	50	0.008	59	0.008	55	0.008											CDV

FREEVIEW DIGITAL TRANSMITTER SITES

Transmitter	NGR	PSB1 BBC A		PSB2 D3&4		PSB3 BBC B		COM4 SDN		COM5 Arqiva A		COM6 Arqiva B		COM7 Arqiva C		COM8 Arqiva D		Pol
		Ch	kW	Ch	kW	Ch	kW	Ch	kW	Ch	kW	Ch	kW	Ch	kW	Ch	kW	
Hemdean	SU710762	52	0.032	46	0.032	48	0.032											CDVH
Hurstbourne Tarrant	SU377533	28	0.002	25	0.002	22	0.002											AV
Lambourn	SU333794	49	0.002	58	0.002	54	0.002											CDV
The Bournes	SU842451	57	0.007	60	0.007	53	0.007											CDV
Tidworth	SU228488	29	0.003	26	0.003	23	0.003											AV
Hastings	TQ806100	25	1	28	1	22	1	23	1	26	1	30	1					AV
Hemel Hempstead	TL088045	44	2	41	2	47	2	50	2	59	2	55	2					WV
Heathfield	TQ566220	52	20	49	20	47	20	42	20	44	20	41	20					BH
Bexhill	TQ729075	46	0.4	40	0.4	43	0.4											BH
Eastbourne	TV606977	26	0.1	30	0.1	23	0.1											AV
Eastbourne Old Town	TV599976	46	0.016	40	0.016	43	0.016											BV
East Dean	TV563988	59	0.008	55	0.008	50	0.008											CDV
Ham Street	TR005332	26	0.002	30	0.002	23	0.002											AV
Hastings Old Town	TQ826101	50	0.02	59	0.02	55	0.02											CDV
Haywards Heath	TQ333234	45	0.007	46	0.007	43	0.007											BV
Hollington Park	TQ795109	50	0.005	59	0.005	55	0.005											CDV
Lamberhurst	TQ673366	60	0.002	54	0.002	57	0.002											CDV
Lewes	TQ409099	28	0.048	22	0.048	25	0.048											AHV
Mountfield	TQ740193	26	0.002	30	0.002	23	0.002											AV
Newhaven	TQ435006	50	0.4	43	0.4	40	0.4											BV
Sedlescombe	TQ784182	26	0.002	30	0.002	23	0.002											AV
St Marks	TQ582373	57	0.026	60	0.026	53	0.026											CDHV
Wye	TR066472	28	0.006	22	0.006	25	0.006											
Huntshaw Cross	SS527220	50	20	59	20	55	20	48	10	52	10	56	10					CDH
Barnstaple	SS575330	46	0.008	43	0.008	40	0.008											BV
Berrynarbor	SS562468	28	0.002	25	0.002	22	0.002											AV
Braunton	SS494369	45	0.002	39	0.002	42	0.002											BH
Brushford	SS921262	27	0.003	24	0.003	21	0.003											AV
Chagford	SX689890	27	0.002	24	0.002	21	0.002											AV
Chambercombe	SS532475	27	0.01	24	0.01	21	0.01											AV
Combe Martin	SS581461	42	0.07	39	0.07	45	0.07											BV
Great Torrington	SS487183	45	0.002	40	0.002	42	0.002											BV
Hartland	SS253256	60	0.06	53	0.06	57	0.06											CDV
Ilfracombe	SS507464	54	0.056	49	0.112	58	0.056											CDV
Muddiford	SS559383	44	0.002	41	0.002	47	0.002											BV
North Bovey	SX741845	46	0.007	43	0.007	40	0.007											BV
Swimbridge	SS622295	26	0.002	23	0.002	29	0.002											AV
Westward Ho!	SS438288	28	0.032	25	0.032	22	0.032											AV

Television Viewer's Guide

FREEVIEW DIGITAL TRANSMITTER SITES | 127

Transmitter	NGR	PSB1 BBC A		PSB2 D3&4		PSB3 BBC B		COM4 SDN		COM5 Arqiva A		COM6 Arqiva B		COM7 Arqiva C		COM8 Arqiva D		Pol
		Ch	kW	Ch	kW	Ch	kW	Ch	kW	Ch	kW	Ch	kW	Ch	kW	Ch	kW	
Woolacombe	SS465428	47	0.002	41	0.002	44	0.002											BV
Idle	SE164374	24	0.05	21	0.05	27	0.05	42	0.05	45	0.05	39	0.05					WV
Keelylang Hill	HY378102	46	20	43	20	40	20	42	10	45	10	39	10					BH
Burgar Hill	HY341261	27	0.004	24	0.004	21	0.004											AHV
Pierowall	HY447497	26	0.01	23	0.01	29	0.01											AV
Keighley	SE069444	49	2	58	2	54	2	57	2	53	2	60	2					CDV
Kilvey Hill	SS672940	23	2	26	2	29	2	25	2	22	2	28	2					AV
Alltwen	SN716039	43	0.002	46	0.002	40	0.002											BV
Briton Ferry	SS747956	43	0.004	46	0.004	40	0.004											BV
Burry Port	SN449019	49	0.002	58	0.004	54	0.004											CDV
Cilfrew	SS775986	39	0.003	45	0.003	42	0.003											BV
Craig-Cefn-Parc	SN682020	43	0.002	46	0.002	40	0.002											BV
Efail-fach	SS786958	39	0.002	45	0.002	42	0.002											BV
Mynydd Emroch	SS775901	43	0.018	46	0.018	40	0.018											BV
Neath Abbey	SS731980	52	0.01	48	0.01	56	0.01											CDH+V
Pontardawe	SN732037	49	0.025	58	0.025	54	0.025											CDV
Knock More	NJ321497	26	20	23	20	29	20	53	10	57	10	60	10					WH
Aviemore	NH940126	28	0.003	25	0.003	22	0.003											AV
Avoch	NH702555	58	0.002	54	0.002	49	0.002											CDV
Balblair Wood	NH593952	50	0.017	59	0.017	55	0.017											CDV
Craigellachie	NJ262443	58	0.014	49	0.014	54	0.014											CDV
Grantown	NJ003267	44	0.07	41	0.07	47	0.07											BV
Kingussie	NN769985	46	0.02	43	0.02	40	0.02											BV
Lairg	NC574056	44	0.003	41	0.003	47	0.003											BV
Lancaster	SD490662	27	2	24	2	21	2	25	2	28	2	22	2					AV
Lark Stoke	SP187426	26	1.26	23	1.26	30	1.26	41	1.26	44	1.26	47	1.26					KV
Limavady	IC711296	50	20	59	20	55	20	54	10	58	10	49	10					CDH
Ballintoy	ID037446	52	0.004	51	0.004	56	0.004											CDV
Ballycastle	ID118397	45	0.003	42	0.003	39	0.003											BV
Buckna	ID222062	44	0.02	41	0.02	47	0.02											BV
Bushmills	IC947405	44	0.002	41	0.002	47	0.002											BV
Castlederg	IH267830	54	0.002	58	0.002	49	0.002											CDV
Claudy	IC561069	57	0.023	53	0.023	60	0.023											CDV
Glenelly Valley	IH576913	26	0.003	29	0.003	23	0.003											AV

Television Viewer's Guide

FREEVIEW DIGITAL TRANSMITTER SITES

Transmitter	NGR	PSB1 BBC A Ch	kW	PSB2 D3&4 Ch	kW	PSB3 BBC B Ch	kW	COM4 SDN Ch	kW	COM5 Arqiva A Ch	kW	COM6 Arqiva B Ch	kW	COM7 Arqiva C Ch	kW	COM8 Arqiva D Ch	kW	Pol
Gortnageeragh	ID155148	45	0.032	42	0.032	39	0.032											BV
Gortnalee	IG955554	28	0.03	25	0.03	22	0.03											AV
Londonderry	IC404175	44	2	41	2	47	2											BV
Muldonagh	IC599100	27	0.02	24	0.02	21	0.02											AV
Plumbridge	IH490906	54	0.005	58	0.005	49	0.005											CDV
Strabane	IH393947	45	0.4	42	0.4	39	0.4											BV
Llanddona	SH583810	57	20	60	20	53	20	43	10	46	10	40	10					WH
Amlwch	SH436920	22	0.007	25	0.007	28	0.007											AV
Arfon	SH476493	41	2	44	2	47	2											BV
Bethesda	SH613663	57	0.005	60	0.005	53	0.005											CDV
Bethesda North	SH627672	28	0.004	25	0.004	22	0.004											AV
Betws-y-Coed	SH825582	27	0.1	24	0.1	21	0.1											AV
Caergybi	SH247817	21	0.002	24	0.002	27	0.002											AV
Caernarfon	SH486629	24	0.005	27	0.005	21	0.005											AV
Cemaes	SH373926	45	0.011	49	0.011	42	0.011											BV
Coed Derw	SH794572	43	0.004	40	0.004	46	0.004											BV
Conway	SH781765	47	0.4	41	0.4	44	0.4											BV
Deiniolen	SH576621	22	0.008	25	0.008	28	0.008											AV
Dolwyddelan	SH740528	44	0.002	41	0.002	47	0.002											BV
Ffestiniog	SH709392	25	0.24	22	0.24	28	0.24											AV
Gronant	SJ088833	26	0.008	29	0.008	23	0.008											AV
Llandecwyn	SH644371	49	0.06	58	0.06	54	0.06											CDV
Llanengan	SH283278	49	0.01	58	0.01	54	0.01											CDH
Maentwrog	SH656406	43	0.006	50	0.006	46	0.006											BV
Mochdre	SH829786	26	0.002	23	0.002	29	0.002											AV
Morfa Nefyn	SH285358	25	0.028	28	0.028	22	0.028											AV
Prestatyn	SJ073822	22	0.002	25	0.002	28	0.002											AV
Trefor	SH376455	45	0.01	49	0.01	42	0.01											BV
Waunfawr	SH529600	25	0.01	28	0.01	22	0.01											AV
Long Mountain	SJ265057	60	0.4	53	0.4	57	0.4											CDV
Broneirion	SO018884	29	0.002	23	0.002	26	0.002											AV
Carno	SN950961	24	0.002	27	0.002	21	0.002											AV
Castle Caereinion	SJ177058	46	0.002	43	0.002	50	0.002											BV
Kerry	SO150908	27	0.0034	24	0.0034	21	0.0034											AV
Llanbrynmair	SH892046	25	0.004	28	0.004	22	0.004											AV
Llandinam	SO050877	41	0.05	47	0.05	44	0.05											BV
Llanfyllin	SJ150180	28	0.025	25	0.025	22	0.025											AV
Llangadfan	SJ020092	28	0.002	25	0.002	22	0.002											AV
Llangurig	SN900794	23	0.002	29	0.002	26	0.002											AV
Llangynog	SJ051259	50	0.002	55	0.002	59	0.002											CDV

FREEVIEW DIGITAL TRANSMITTER SITES 129

Transmitter	NGR	PSB1 BBC A		PSB2 D3&4		PSB3 BBC B		COM4 SDN		COM5 Arqiva A		COM6 Arqiva B		COM7 Arqiva C		COM8 Arqiva D		Pol
		Ch	kW	Ch	kW	Ch	kW	Ch	kW	Ch	kW	Ch	kW	Ch	kW	Ch	kW	
Llanidloes	SN947843	25	0.002	28	0.002	22	0.002											AV
Llanrhaeadr-ym-Mochnant	SJ174260	45	0.015	49	0.015	42	0.015											BV
Moel-y-Sant	SJ151105	27	0.023	24	0.023	21	0.023											AV
Tregynon	SO110963	28	0.007	25	0.007	22	0.007											AV
Malvern	**SO774464**	**53**	**0.4**	**57**	**0.4**	**60**	**0.4**	**50**	**0.4**	**59**	**0.4**	**55**	**0.4**					**CDV**
Mendip	**ST563488**	**49**	**100**	**54**	**100**	**58**	**100**	**48**	**100**	**56**	**100**	**52**	**100**	**33**	**72**	**35**	**69**	**WH**
Avening	ST880975	41	0.002	47	0.002	44	0.002											BV
Backwell	ST498716	25	0.019	28	0.019	22	0.019											AV
Bath	ST769654	25	0.05	28	0.05	22	0.05											AV
Blakeney	SO664069	24	0.025	27	0.025	21	0.025											AV
Box	ST832688	43	0.002	40	0.002	46	0.002											BV
Bristol Barton House	ST611729	26	0.018	23	0.018	29	0.018											AH
Bristol Montpelier	ST590745	27	0.01	24	0.01	21	0.01											AV
Bristol Warmley	ST654736	45	0.002	42	0.002	39	0.002											EV
Bruton	ST680341	43	0.002	40	0.002	46	0.002											BV
Burrington	ST477606	60	0.032	53	0.032	57	0.032											CDH
Calne	ST997699	24	0.02	27	0.02	21	0.02											AV
Carhampton	ST016426	24	0.02	27	0.02	21	0.02											AV
Cerne Abbas	ST645012	29	0.08	26	0.08	23	0.08											AV
Chalford	SO883017	24	0.025	27	0.025	21	0.025											AV
Chalford Vale	SO886024	43	0.003	46	0.003	40	0.003											EV
Chilfrome	SY580991	43	0.013	40	0.013	46	0.013											BH
Chiseldon	SU190801	27	0.016	24	0.016	21	0.016											AV
Chitterne	ST995441	43	0.002	40	0.002	46	0.002											BV
Cirencester	SP004057	23	0.1	29	0.1	26	0.1											AV
Clearwell	SO574084	59	0.002	50	0.002	55	0.002											CDV
Coleford	SO569107	45	0.002	39	0.002	42	0.002											BV
Corsham	ST869699	40	0.008	46	0.008	43	0.008											BVH
Countisbury	SS749501	59	0.042	50	0.042	55	0.042											CDH
Crewkerne	ST444092	43	0.002	40	0.002	46	0.002											BV
Crockerton	ST877428	43	0.011	46	0.011	40	0.011											BV
Dursley	ST788963	43	0.011	40	0.011	46	0.011											BV
Easter Compton	ST567830	53	0.0025	60	0.0025	57	0.0025											CDV
Exford	SS847379	41	0.002	47	0.002	44	0.002											BV
Frome	ST778482	26	0.002	23	0.002	29	0.002											AV
Hutton	ST361588	59	0.04	50	0.04	55	0.04											EV
Kewstoke	ST347639	43	0.016	40	0.016	46	0.016											BV
Kilve	ST143425	59	0.003	50	0.003	55	0.003											CDH
Lydbrook	SO600163	43	0.002	40	0.002	46	0.002											BV
Marlborough	SU209688	25	0.02	28	0.02	22	0.02											AV

Television Viewer's Guide

FREEVIEW DIGITAL TRANSMITTER SITES

Transmitter	NGR	PSB1 BBC A		PSB2 D3&4		PSB3 BBC B		COM4 SDN		COM5 Arqiva A		COM6 Arqiva B		COM7 Arqiva C		COM8 Arqiva D		Pol
		Ch	kW	Ch	kW	Ch	kW	Ch	kW	Ch	kW	Ch	kW	Ch	kW	Ch	kW	
Monksilver	ST085362	40	0.015	43	0.015	46	0.015											BV
Nailsworth	ST849990	23	0.006	26	0.006	29	0.006											AV
Ogbourne St. George	SU205732	43	0.0026	40	0.0026	46	0.0026											BV
Parkend	SO616083	41	0.002	47	0.002	44	0.002											BV
Pillowell	SO625065	43	0.002	40	0.002	46	0.002											BVH
Porlock	SS883462	24	0.005	27	0.005	21	0.005											BV
Portbury	ST502751	24	0.002	27	0.002	21	0.002											AV
Portishead	ST458764	59	0.0074	50	0.0074	55	0.0074											EV
Redbrook	SO538092	42	0.002	45	0.002	39	0.002											BV
Redcliff Bay	ST439751	53	0.01	60	0.01	57	0.01											CDH
Roadwater	ST026375	23	0.006	26	0.006	29	0.006											AH
Seagry Court	SU149881	41	0.02	47	0.02	44	0.02											BV
Siston	ST668744	26	0.008	23	0.008	29	0.008											AV
Slad	SO872055	23	0.002	29	0.002	26	0.002											AH
Stroud	SO836077	40	0.2	43	0.2	46	0.2											BV
Tintern	SO538002	24	0.002	27	0.002	21	0.002											AV
Ubley	ST538594	24	0.016	27	0.016	21	0.016											AV
Upavon	SU145518	23	0.028	29	0.028	26	0.028											AV
Washford	ST058410	59	0.012	50	0.012	55	0.012											CDV
West Lavington	ST999525	23	0.0024	26	0.0024	29	0.0024											AV
Westwood	ST817597	43	0.02	40	0.02	46	0.02											BV
Woodcombe	SS951458	59	0.01	50	0.01	55	0.01											CDV
Wootton Courtenay	SS934426	25	0.022	28	0.022	22	0.022											AV
Midhurst	SU912250	55	20	56	20	58	20	54	10	59	10	50	10					CDH
Haslemere	SU886331	28	0.01	22	0.01	25	0.01											AV
Steyning	TQ184121	46	0.056	40	0.056	43	0.056											BV
Moel-y-Parc	SJ123702	45	20	39	20	42	20	51	20	52	20	48	20	32	14	34	14	WH
Bala	SH969375	26	0.04	23	0.04	29	0.04											AV
Betws-yn-Rhos	SH899756	27	0.003	24	0.003	21	0.003											AV
Cefn-Mawr	SJ267409	57	0.025	53	0.025	60	0.05											CDV
Cerrigydrudion	SH933482	26	0.006	23	0.006	29	0.006											AV
Corwen	SJ080431	28	0.04	25	0.04	22	0.04											AV
Cyffylliog	SJ063580	28	0.002	25	0.002	22	0.002											AV
Glyn Ceiriog	SJ203386	58	0.002	49	0.002	54	0.002											CDV
Glyndyfrdwy	SJ158429	50	0.002	59	0.002	55	0.002											CDV
Llanarmon-yn-Ial	SJ194582	27	0.002	24	0.002	21	0.002											AV
Llandderfel	SH990359	50	0.002	55	0.002	59	0.002											CDV
Llanddulas	SH910784	26	0.003	23	0.003	29	0.003											AH
Llangernyw	SH881660	28	0.002	22	0.002	25	0.002											AV
Llangollen	SJ204421	58	0.002	49	0.002	54	0.002											CDV

Television Viewer's Guide

FREEVIEW DIGITAL TRANSMITTER SITES

Transmitter	NGR	PSB1 BBC A		PSB2 D3&4		PSB3 BBC B		COM4 SDN		COM5 Arqiva A		COM6 Arqiva B		COM7 Arqiva C		COM8 Arqiva D		Pol
		Ch	kW	Ch	kW	Ch	kW	Ch	kW	Ch	kW	Ch	kW	Ch	kW	Ch	kW	
Llanuwchllyn	SH873326	46	0.006	43	0.006	40	0.006											BV
Pen-y-Banc	SJ077497	27	0.002	24	0.002	21	0.002											AV
Penmaen Rhos	SH877779	22	0.028	25	0.028	28	0.028											AV
Pontfadog	SJ219356	28	0.002	25	0.002	22	0.002											AV
Pwll-glas	SJ119541	26	0.002	23	0.002	29	0.002											AV
Storeton Wales	SJ314841	57	0.5	53	0.5	60	0.5											CDH
Wrexham-Rhos	SJ301537	28	0.08	22	0.08	25	0.08											AV
Nottingham	**SK503435**	27	0.4	24	0.4	21	0.4	51	0.4	52	0.4	48	0.4					WV
Oliver's Mount	**TA040869**	57	2	60	2	53	2	54	1	58	1	49	1					CDV
Hunmanby	TA092779	48	0.025	39	0.025	42	0.025											BV
Oxford	**SP567105**	53	100	60	100	57	100	50	50	59	50	55	50	31	16	37	17	WH
Ascott under Wychwood	SP287193	27	0.006	24	0.006	21	0.006											AV
Charlbury	SP344197	44	0.01	41	0.01	47	0.01											BV
Guiting Power	SP101233	44	0.005	41	0.005	47	0.005											BV
Icomb Hill	SP201228	28	0.022	25	0.022	22	0.022											AV
Over Norton	SP309282	48	0.031	56	0.031	52	0.031											CDV
Pendle Forest	**SD825384**	28	0.1	25	0.1	22	0.1	27	0.1	21	0.1	24	0.1					AV
Plympton	**SX531555**	54	0.4	49	0.4	58	0.4	42	0.4	45	0.4	56	0.4					EV
Pontop Pike	**NZ148526**	58	100	54	100	49	100	50	50	59	50	55	50	33	34	34	34	WH
Allenheads	NY840469	27	0.002	24	0.002	21	0.002											AV
Alston	NY730478	39	0.08	42	0.08	45	0.08											BV
Bellingham	NY832812	27	0.01	24	0.01	21	0.01											AV
Blaydon Burn	NZ167627	44	0.002	41	0.002	47	0.002											BV
Byrness	NT765016	27	0.02	24	0.02	21	0.02											AV
Catton Beacon	NY822590	40	0.028	43	0.028	46	0.028											BV
Durham	NZ264423	41	0.003	44	0.003	47	0.003											BV
Esh	NZ198444	39	0.002	42	0.002	45	0.002											BV
Falstone	NY723863	44	0.002	47	0.002	41	0.002											BV
Felling	NZ276616	40	0.002	46	0.002	43	0.002											BV
Haydon Bridge	NY809630	44	0.02	47	0.02	41	0.02											BV
Hedleyhope	NZ162397	41	0.004	44	0.004	47	0.004											BH
Humshaugh	NY906711	39	0.012	42	0.012	45	0.012											BV
Ireshopeburn	NY862381	55	0.002	50	0.002	59	0.002											CDV
Kielder	NY625969	26	0.005	29	0.005	23	0.005											AV
Morpeth	NZ218864	22	0.009	25	0.009	28	0.009											AV
Newton	NZ035653	26	0.4	29	0.4	23	0.4											AV

Television Viewer's Guide

FREEVIEW DIGITAL TRANSMITTER SITES

Transmitter	NGR	PSB1 BBC A		PSB2 D3&4		PSB3 BBC B		COM4 SDN		COM5 Arqiva A		COM6 Arqiva B		COM7 Arqiva C		COM8 Arqiva D		Pol
		Ch	kW	Ch	kW	Ch	kW	Ch	kW	Ch	kW	Ch	kW	Ch	kW	Ch	kW	
Seaham	NZ402485	44	0.1	47	0.1	41	0.1											BV
Shotleyfield	NZ065532	22	0.04	25	0.04	28	0.04											AV
Staithes	NZ781189	52	0.002	51	0.002	48	0.002											BV
Sunderland	NZ391547	52	0.006	51	0.006	48	0.006											BV
Wall	NY909676	52	0.004	51	0.004	48	0.004											BH
Weardale	NZ025384	44	0.2	47	0.2	41	0.2											BV
Whitaside	SD979964	44	0.003	47	0.003	41	0.003											BV
Pontypool	ST284990	23	0.05	26	0.05	29	0.05	25	0.05	22	0.05	28	0.05					AV
Presely	SN172306	43	20	46	20	40	20	42	10	45	10	39	10					BH
Abergwynfi	SS886971	24	0.002	21	0.002	27	0.002											AV
Broad Haven	SM861130	51	0.016	58	0.016	54	0.016											CDV
Bronnant	SN664664	23	0.003	29	0.003	26	0.003											AV
Croeserw	SS858952	49	0.024	58	0.024	54	0.024											CDHV
Cynwyl Elfed	SN375273	25	0.002	22	0.002	28	0.002											AV
Dolgellau	SH727185	59	0.004	55	0.004	50	0.004											CDV
Duffryn	SS834956	25	0.002	22	0.002	28	0.002											AV
Ferryside	SN371104	21	0.025	30	0.025	24	0.025	27	0.005									AV
Fishguard	SM944391	49	0.025	58	0.025	54	0.025											CDV
Glyncorrwg	SS871989	39	0.002	45	0.002	42	0.002											BV
Haverfordwest	SN028261	56	0.2	52	0.2	47	0.2											CDH
Llandyfriog	SN348412	25	0.022	22	0.022	28	0.022											AV
Llandysul	SN425409	60	0.015	57	0.015	53	0.015											CDV
Llangranog	SN322538	25	0.008	22	0.008	28	0.008											AV
Llangybi	SN614524	25	0.003	22	0.003	28	0.003											AV
Llwyn Onn	SH625175	25	0.014	22	0.014	28	0.014											AV
Mynydd Pencarreg	SN577430	49	0.024	58	0.024	54	0.024											CDV
Newport Bay	SN066414	47	0.01	41	0.01	44	0.01											BV
Pembroke Dock	SM967028	49	0.01	58	0.01	54	0.01											CDV
Pencader	SN451370	23	0.002	26	0.002	29	0.002											AV
Rheola	SN841061	59	0.02	55	0.02	50	0.02											CDV
St.Davids	SM751254	23	0.016	26	0.016	29	0.016											AV
St.Dogmaels	SN165452	23	0.003	26	0.003	29	0.003											AV
Trefin	SM848310	25	0.1	22	0.1	28	0.1											AV
Tregaron	SN686605	55	0.003	50	0.003	59	0.003											CDV
Ystumtuen	SN740795	39	0.02	45	0.02	42	0.02											BV
Redruth	SW690395	44	20	41	20	47	20	48	10	52	10	51	10					BH
Alverton	SW457299	27	0.002	24	0.002	21	0.002											AV
Boscastle	SX096911	25	0.011	22	0.011	28	0.011											AV
Bossiney	SX067889	46	0.04	43	0.04	50	0.04											BV

Television Viewer's Guide

FREEVIEW DIGITAL TRANSMITTER SITES

Transmitter	NGR	PSB1 BBC A		PSB2 D3&4		PSB3 BBC B		COM4 SDN		COM5 Arqiva A		COM6 Arqiva B		COM7 Arqiva C		COM8 Arqiva D		Pol
		Ch	kW	Ch	kW	Ch	kW	Ch	kW	Ch	kW	Ch	kW	Ch	kW	Ch	kW	
Gulval	SW475315	25	0.005	23	0.005	29	0.005											AV
Helston	SW651275	54	0.002	49	0.002	58	0.002											CDV
Isles of Scilly	SV911124	27	0.1	24	0.1	21	0.1											AV
Mevagissey	SX011445	46	0.01	43	0.01	50	0.01											BH
Penryn	SW787334	50	0.011	59	0.011	55	0.011											CDV
Perranporth	SW758533	46	0.016	43	0.016	50	0.016											BV
Porthleven	SW626257	26	0.002	23	0.002	29	0.002											AH
Porthtowan	SW694478	27	0.016	24	0.016	21	0.016											AV
Portreath	SW658455	26	0.02	23	0.02	29	0.02											AV
Praa Sands	SW572284	50	0.025	59	0.025	56	0.025											CDV
St.Anthony-in-Roseland	SW852318	26	0.002	23	0.002	29	0.002											AV
St Austell	SX008535	50	0.25	59	0.25	56	0.25											CDV
St.Just	SW382331	56	0.25	50	0.25	53	0.25											CDV
Truro	SW835442	54	0.004	49	0.004	58	0.004											CDV
Reigate	TQ257521	60	2	57	2	53	2	21	2	24	2	27	2					WV
Ridge Hill	SO630333	28	20	25	20	22	20	21	10	24	10	27	10	32	10	34	10	AH
Ridge Hill	SO630333			29	20			Carries West Region ITV-1 on SE beam only										
Andoversford	SP002183	50	0.011	59	0.011	55	0.011											CDV
Eardiston	SO706682	54	0.002	49	0.002	58	0.002											CDV
Ewyas Harold	SO389269	44	0.002	41	0.002	47	0.002											BV
Garth Hill	SO273726	53	0.007	60	0.007	57	0.007											CDHV
Hazler Hill	SO464928	45	0.005	42	0.005	39	0.005											BV
Hereford	SO524364	44	0.007	41	0.007	47	0.007											BV
Hope-under-Dinmore	SO504525	57	0.002	60	0.002	53	0.002											CDV
Kington	SO290553	44	0.02	41	0.02	47	0.02											BV
Knucklas	SO270747	39	0.002	42	0.002	45	0.002											BV
Ludlow	SO498741	45	0.005	42	0.005	39	0.005											BV
New Radnor	SO269623	44	0.025	41	0.025	47	0.025											BV
Oakeley Mynd	SO346875	45	0.01	49	0.01	42	0.01											BV
Peterchurch	SO360380	53	0.015	60	0.015	57	0.015											CDV
Presteigne	SO337661	45	0.003	42	0.003	49	0.003											BV
Ross-on-Wye	SO605243	50	0.002	59	0.002	55	0.002											CDV
St.Briavels	SO557049	46	0.005	43	0.005	40	0.005											BV
Upper Soudley	SO662091	46	0.002	43	0.002	50	0.002											BV
Rosemarkie	NH762623	45	20	39	20	42	20	43	10	46	10	40	10					BH
Auchmore Wood	NH484502	28	0.02	25	0.02	22	0.02											AV
Cromarty	NH787676	28	0.002	25	0.002	22	0.002											AV
Fodderty	NH512606	58	0.024	49	0.024	54	0.024											CDV
Fort Augustus	NH361049	26	0.002	23	0.002	29	0.002											AV

FREEVIEW DIGITAL TRANSMITTER SITES

Transmitter	NGR	PSB1 BBC A		PSB2 D3&4		PSB3 BBC B		COM4 SDN		COM5 Arqiva A		COM6 Arqiva B		COM7 Arqiva C		COM8 Arqiva D		Pol
		Ch	kW	Ch	kW	Ch	kW	Ch	kW	Ch	kW	Ch	kW	Ch	kW	Ch	kW	
Glen Convinth	NH506393	27	0.007	24	0.007	21	0.007											AV
Glen Urquhart	NH442295	44	0.02	41	0.02	47	0.02											BVH
Inverness	NH667447	50	0.007	55	0.007	59	0.007											CDV
Tomatin	NH822288	28	0.003	25	0.003	22	0.003											AV
Tomich	NH306276	27	0.003	24	0.003	21	0.003											AV
Tomich Link	NH321265	45	0.002	39	0.002	42	0.002											BV
Wester Erchite	NH577307	27	0.003	24	0.003	21	0.003											AV
Rosneath	NS258811	49	2	58	2	54	2	53	2	57	2	60	2					CDV
Ardnadam	NS167799	42	0.014	45	0.014	39	0.014											BV
Garelochhead	NS235918	42	0.002	45	0.002	39	0.002											BV
Rowridge (Horiz)	SZ447865	24	200	27	200	21	200	25	50	22	50	28	50	31	23	37	25	AH
Rowridge (Vert)	SZ447865	24	200	27	200	21	200	25	200	22	200	28	200	31	0	37	0	AV
Bovington	SY844878	44	0.002	41	0.002	47	0.002											BV
Brading	SZ612872	49	0.005	46	0.005	43	0.005											BV
Brighstone	SZ435816	44	0.07	41	0.07	47	0.07											BV
Canford Heath	SZ036939	45	0.002	46	0.002	42	0.002											BV
Cheselbourne	SY768985	57	0.002	53	0.002	60	0.002											CDV
Corfe Castle	SY972821	44	0.004	41	0.004	47	0.004											BV
Donhead	ST907230	44	0.006	41	0.006	47	0.006											BV
Findon	TQ120072	44	2	41	2	47	2											BV
Horndean	SU698152	56	0.003	50	0.003	40	0.003											CDV
Luccombe	SZ581800	54	0.02	59	0.02	50	0.02											CDV
Lulworth	SY824815	50	0.002	59	0.002	55	0.002											CDV
Luscombe Valley	SZ047908	45	0.002	46	0.002	42	0.002											BV
Millbrook	SU374138	44	0.008	41	0.008	47	0.008											BVH
Piddletrenthide	SY704989	45	0.011	39	0.011	42	0.011											BVH
Poole	SZ037921	57	0.02	60	0.02	53	0.02											CDV
Poulner	SU145053	44	0.01	41	0.01	47	0.01											BH
Shrewton	SU072438	44	0.002	41	0.002	47	0.002											BV
Singleton	SU866131	44	0.003	41	0.003	47	0.003											BV
Sutton Row	ST976284	29	0.05	23	0.05	26	0.05											AV
Till Valley	SU066370	46	0.015	43	0.015	50	0.015											BV
Ventnor	SZ567783	48	0.4	49	0.4	52	0.4											CDV
Westbourne	SZ067916	44	0.008	41	0.008	47	0.008											BV
Winterbourne Steepleton	SY629893	39	0.002	45	0.002	42	0.002											BV
Winterbourne Stickland	ST838051	46	0.4	43	0.4	40	0.4											BV
Rumster Forest	ND197385	27	20	24	20	21	20	30	10	59	10	55	10					WH
Ben Tongue	NC604588	45	0.007	49	0.007	42	0.007											BV
Durness	NC409672	57	0.002	53	0.002	60	0.002											CDV

Television Viewer's Guide

FREEVIEW DIGITAL TRANSMITTER SITES | 135

Transmitter	NGR	PSB1 BBC A		PSB2 D3&4		PSB3 BBC B		COM4 SDN		COM5 Arqiva A		COM6 Arqiva B		COM7 Arqiva C		COM8 Arqiva D		Pol
		Ch	kW	Ch	kW	Ch	kW	Ch	kW	Ch	kW	Ch	kW	Ch	kW	Ch	kW	
Melvich	NC880637	44	0.011	41	0.011	47	0.011											BV
Thurso	ND119673	57	0.002	60	0.002	53	0.002											CDV
Saddleworth	SD987050	45	0.4	49	0.4	42	0.4	51	0.4	52	0.4	48	0.4					BV
Sailsbury	SU136285	57	2	60	2	53	2	50	2	59	2	55	2					CDV
Sandy Heath	TL204494	27	180	24	180	21	180	51	170	52	170	48	170	32	50	34	50	WH
Dallington Park	SP742612	50	0.04	59	0.04	55	0.04											CDV
Kimpton	TL173178	57	0.002	53	0.002	60	0.002											CDV
Luton	TL081210	50	0.08	59	0.08	55	0.08											CDV
Selkirk	NT500294	50	10	59	10	55	10	57	5	53	5	60	5					CDH
Bonchester Bridge	NT589114	45	0.002	39	0.002	42	0.002											BV
Clovenfords	NT464358	27	0.002	24	0.002	21	0.002											AV
Eyemouth	NT947599	21	0.4	24	0.4	27	0.4											AV
Galashiels	NT507360	44	0.02	41	0.02	47	0.02											AV
Hawick	NT509147	26	0.01	23	0.01	29	0.01											AV
Innerleithen	NT325368	49	0.016	54	0.016	58	0.016											CDV
Jedburgh	NT661224	44	0.032	41	0.032	47	0.032											BV
Lauder	NT506502	28	0.0028	25	0.0028	22	0.0028											AV
Peebles	NT228416	28	0.02	25	0.02	22	0.02											AV
Stow	NT448445	26	0.002	23	0.002	29	0.002											AV
Yetholm	NT836283	44	0.002	41	0.002	47	0.002											BV
Sheffield (Crosspool)	SK324870	27	1	24	1	21	1	42	1	45	1	39	1	31	1	37	1	WV
Stockland Hill	ST222014	26	50	23	50	29	50	25	25	22	25	28	25					AH
Bampton	SS967237	39	0.006	45	0.006	42	0.006											BV
Beaminster	ST490024	50	0.015	59	0.015	55	0.015											CDV
Beer	SY230896	50	0.005	59	0.005	56	0.005											CDV
Bincombe Hill	SY687848	50	0.006	59	0.006	56	0.006											CDV
Branscombe	SY195882	44	0.002	47	0.002	41	0.002											BV
Bridport	SY453915	44	0.017	41	0.017	47	0.017											BV
Budleigh Salterton	SY043825	60	1	53	1	57	1											CDV
Charmouth	SY373933	47	0.1	44	0.1	41	0.1											BV
Chideock	SY412933	56	0.02	52	0.02	48	0.02											CDHV
Crediton	SS825009	46	0.007	43	0.007	40	0.007											BV
Culm Valley	ST108148	45	0.012	39	0.012	42	0.012											BV
Dawlish	SX950772	48	0.02	52	0.02	56	0.02											CDV
Dunsford	SX811880	42	0.002	45	0.002	49	0.002											BV

Television Viewer's Guide

FREEVIEW DIGITAL TRANSMITTER SITES

Transmitter	NGR	PSB1 BBC A		PSB2 D3&4		PSB3 BBC B		COM4 SDN		COM5 Arqiva A		COM6 Arqiva B		COM7 Arqiva C		COM8 Arqiva D		Pol
		Ch	kW	Ch	kW	Ch	kW	Ch	kW	Ch	kW	Ch	kW	Ch	kW	Ch	kW	
Exeter St.Thomas	SX898922	44	0.05	41	0.05	47	0.05											BV
Honiton	SY164997	42	0.002	49	0.002	45	0.002											BV
Pennsylvania	SX934948	54	0.002	56	0.002	58	0.002											CDV
Preston	SY707833	52	0.02	58	0.02	54	0.02											CDV
Rampisham	ST544008	46	0.002	43	0.002	40	0.002											BH
Stokeinteignhead	SX909711	44	0.002	41	0.002	47	0.002											BV
Tiverton	SS939126	46	0.018	43	0.018	50	0.018											BV
Weymouth	SY663778	47	0.4	44	0.4	41	0.4											BV
Storeton	SJ314841	28	0.56	25	0.56	22	0.56	23	0.56	26	0.56	29	0.56					AV
Sudbury	TL913379	44	100	41	100	47	100	58	100	60	100	56	100					EH
Burnham on Crouch	TL940966	49	0.4	52	0.4	42	0.4											BH
Clacton	TM183163	49	0.4	52	0.4	42	0.4											BH
Felixstowe	TM305346	45	0.2	51	0.2	53	0.2											BV
Ipswich Stoke	TM159443	26	0.028	25	0.028	29	0.028											AV
Rouncefall	TQ856931	44	4	41	4	47	4											BH
Somersham	TM086492	28	0.002	25	0.002	22	0.002											AV
Wivenhoe Park	TM027240	57	0.01	54	0.01	49	0.01											CDV
Woodbridge	TM271498	48	0.1	52	0.1	54	0.1											CDV
Sutton Coldfield	SK113003	43	200	46	200	40	200	42	200	45	200	39	200	33	89	35	86	WH
Allesley Park	SP296796	25	0.007	22	0.007	28	0.007											AV
Brailes	SP319379	24	0.008	21	0.008	27	0.008											WV
Bretch Hill	SP438400	56	0.087	48	0.087	52	0.087											CDV
Bridgnorth	SO719913	59	0.003	50	0.003	55	0.003											CDV
Cheadle	SK030435	56	0.005	48	0.005	52	0.005											CDV
Earl Sterndale	SK090666	49	0.04	58	0.04	54	0.04											CDV
Edgbaston	SP058851	24	0.004	21	0.004	27	0.004											AV
Gib Heath	SP056883	50	0.003	59	0.003	55	0.003											CDV
Gravelly Hill	SP108897	50	0.003	59	0.003	55	0.003											CDH
Haden Hill	SO967846	48	0.016	52	0.016	56	0.016											BV
Hamstead	SP043931	24	0.002	21	0.002	27	0.002											AV
Harborne	SP017836	56	0.04	52	0.04	48	0.04											CDV
Hartington	SK117601	56	0.007	48	0.007	52	0.007											CDV
Ipstones Edge	SK043506	60	0.006	57	0.006	53	0.006											CDV
Ironbridge	SJ678032	49	0.002	58	0.002	54	0.002											CDV
Kenilworth	SP298726	60	0.01	57	0.01	53	0.01											CDV
Kidderminster	SO808739	49	0.4	58	0.4	54	0.4											CDV
Kinver	SO855831	48	0.005	56	0.005	52	0.005											CDV
Leamington Spa	SP329663	59	0.04	50	0.04	55	0.04											CDV
Leek	SJ999561	25	0.2	22	0.2	28	0.2											AV

FREEVIEW DIGITAL TRANSMITTER SITES 137

Transmitter	NGR	PSB1 BBC A		PSB2 D3&4		PSB3 BBC B		COM4 SDN		COM5 Arqiva A		COM6 Arqiva B		COM7 Arqiva C		COM8 Arqiva D		Pol
		Ch	kW	Ch	kW	Ch	kW	Ch	kW	Ch	kW	Ch	kW	Ch	kW	Ch	kW	
Long Compton	SP285338	25	0.004	22	0.004	28	0.004											AV
Oakamoor	SK057446	24	0.002	21	0.002	27	0.002											AV
Perry Beeches	SP067932	25	0.002	22	0.002	28	0.002											AV
Queslett	SP063948	49	0.003	58	0.003	54	0.003											CDV
Redditch	SP028683	25	0.002	22	0.002	28	0.002											AV
Repton	SK307261	60	0.008	57	0.008	53	0.008											CDV
Rugeley	SK034179	56	0.008	52	0.008	48	0.008											CDV
Tenbury Wells	SO588691	60	0.003	57	0.003	53	0.003											CDV
Turves Green	SP022784	50	0.002	59	0.002	55	0.002											CDV
Whittingslow	SO429886	60	0.011	57	0.011	53	0.011											CDV
Winchcombe	SP036287	49	0.002	58	0.002	54	0.002											CDV
Winshill	SK272241	60	0.06	57	0.06	53	0.06											CDV
Woodford Halse	SP540530	25	0.002	22	0.002	28	0.002											AV
Tacolneston	**TM131959**	55	100	59	100	50	100	42	100	45	100	39	100	31	27	37	24	WH
Aldeburgh	TM443596	28	10	23	10	25	10											AV
Bramford	TM117458	27	0.004	24	0.004	21	0.004											AV
Burnham	TF795428	41	0.4	47	0.4	44	0.4											BV
Bury St.Edmunds	TL861651	28	0.017	25	0.017	22	0.017											AV
Creake	TF846370	48	0.002	52	0.002	56	0.002											CDV
Gorleston	TG521049	26	0.002	23	0.002	29	0.002											AV
Great Yarmouth	TG521070	47	0.4	41	0.4	44	0.4											BV
King's Lynn	TF677281	40	0.3	46	0.3	43	0.3											BV
Linnet Valley	TL826636	26	0.0032	23	0.0032	29	0.0032											AV
Little Walsingham	TF926365	44	0.011	41	0.011	47	0.011											BV
Lowestoft	TM543938	58	0.4	60	0.4	56	0.4											CDV
Norwich (Central)	TG237098	46	0.0068	43	0.0068	52	0.0068											BV
Overstrand	TG234408	48	0.063	47	0.063	51	0.063											BV
Thetford	TL865839	26	0.008	23	0.008	29	0.008											AV
Wells-next-the-Sea	TF985423	57	1	51	1	56	1											CDV
West Runton	TG186412	26	2	23	2	29	2											AV
The Wrekin	**SJ628082**	26	20	23	20	30	20	41	10	44	10	47	10					EH
Bucknell	SO358733	45	0.002	49	0.002	42	0.002											BV
Clun	SO324799	55	0.011	59	0.011	50	0.011											CDV
Coalbrookdale	SJ671042	50	0.002	43	0.002	46	0.002											BV
Halesowen	SO971826	58	0.002	49	0.002	54	0.002											CDV
Torosay	**NM703358**	28	4	25	4	22	4	23	4	26	2	29	2					AV
Acharacle	NM678695	46	0.002	43	0.002	50	0.002											BV
Arisaig	NM669873	26	0.002	23	0.002	29	0.002											AV
Ballachulish	NN059593	23	0.004	29	0.004	26	0.004											AV

Television Viewer's Guide

FREEVIEW DIGITAL TRANSMITTER SITES

Transmitter	NGR	PSB1 BBC A		PSB2 D3&4		PSB3 BBC B		COM4 SDN		COM5 Arqiva A		COM6 Arqiva B		COM7 Arqiva C		COM8 Arqiva D		Pol
		Ch	kW	Ch	kW	Ch	kW	Ch	kW	Ch	kW	Ch	kW	Ch	kW	Ch	kW	
Bellanoch	NR803919	45	0.01	42	0.01	49	0.01											BV
Castlebay	NL652979	27	0.002	24	0.002	21	0.002											AV
Cow Hill	NN112735	46	0.013	43	0.013	50	0.013											BV
Dalmally	NN143260	44	0.008	41	0.008	47	0.008											BV
Dychliemore	NN138238	28	0.0005	25	0.0005	22	0.0005											AH
Glengorm	NM466565	44	1	41	1	47	1											BV
Kilmelford	NM818101	50	0.015	59	0.015	55	0.015											CDV
Kinlochleven	NN178630	59	0.002	55	0.002	50	0.002											BV
Kintraw	NM830048	46	0.004	43	0.004	40	0.004											CDV
Loch Feochan	NM861256	54	0.008	49	0.008	58	0.008											CDV
Mallaig	NM676965	46	0.004	43	0.004	50	0.004											BV
Oban	NM850289	44	0.002	41	0.002	47	0.002											BV
Onich	NN017618	49	0.003	54	0.003	58	0.003											CDVH
Spean Bridge	NN218819	27	0.014	24	0.014	21	0.014											AV
Strontian	NM830657	39	0.003	45	0.003	42	0.003											BV
Taynuilt	NM994313	46	0.002	43	0.002	50	0.002											BV
Tayvallich	NR740866	46	0.002	43	0.002	50	0.002											BV
Tunbridge Wells	TQ607440	52	4	49	4	47	4	42	4	44	4	41	4					BV
Waltham	SK809233	49	50	54	50	58	50	29	25	56	25	57	25	31	10	37	10	WH
Ambergate	SK351513	25	0.037	28	0.037	22	0.037											AV
Ashbourne	SK182460	25	0.05	28	0.05	22	0.05											AV
Ashford-in-Water	SK189691	23	0.003	26	0.003	30	0.003											AV
Belper	SK337462	59	0.006	50	0.006	55	0.006											CDV
Birchover	SK241616	39	0.005	45	0.005	42	0.005											BH
Bolehill	SK295552	53	0.1	57	0.1	60	0.1											CDV
Darley Dale	SK275642	44	0.003	41	0.003	47	0.003											BV
Derby	SK329342	48	0.04	51	0.04	52	0.04											CDH
Eastwood	SK463470	23	0.002	26	0.002	30	0.002											AV
Leicester City	SK584033	25	0.002	28	0.002	22	0.002											AV
Little Eaton	SK371419	23	0.004	26	0.004	30	0.004											AV
Matlock	SK297589	24	0.003	27	0.003	21	0.003											AV
Parwich	SK185542	24	0.02	27	0.02	21	0.02											AV
Stamford	TF032067	47	0.02	41	0.02	44	0.02											BV
Stanton Moor	SK246637	59	0.4	50	0.4	55	0.4											CDV
Wenvoe	ST110742	41	100	44	100	47	100	42	50	45	50	39	50	31	47	37	52	WH
Aberbeeg	SO215029	43	0.002	50	0.002	46	0.002											BV
Abercynon	ST093952	58	0.002	54	0.002	49	0.002											CDH
Abergavenny	SO244126	39	0.2	42	0.2	45	0.2											BV
Abertillery	SO224023	25	0.056	22	0.056	28	0.056											AV

FREEVIEW DIGITAL TRANSMITTER SITES 139

Transmitter	NGR	PSB1 BBC A		PSB2 D3&4		PSB3 BBC B		COM4 SDN		COM5 Arqiva A		COM6 Arqiva B		COM7 Arqiva C		COM8 Arqiva D		Pol
		Ch	kW	Ch	kW	Ch	kW	Ch	kW	Ch	kW	Ch	kW	Ch	kW	Ch	kW	
Abertridwr	ST123886	60	0.01	57	0.01	53	0.01											CDV
Bargoed	SO145010	24	0.06	21	0.06	27	0.06											AV
Bedlinog	SO102005	24	0.002	21	0.002	27	0.002											AV
Blackmill	SS930867	25	0.002	22	0.002	28	0.002											AV
Blaenau Gwent	SO215049	60	0.0022	57	0.0022	53	0.0022											CDV
Blaenavon	SO277063	60	0.03	57	0.03	53	0.03											CDV
Blaenllechau	SS998977	24	0.002	21	0.002	27	0.002											AH
Blaina	SO196083	43	0.02	40	0.02	46	0.02											BV
Brecon	SO054287	49	0.2	54	0.2	58	0.2											CDV
Chepstow	ST544939	24	0.002	21	0.002	27	0.002											AV
Clydach	SO227125	23	0.002	29	0.002	26	0.002											AV
Clyro	SO204432	41	0.032	47	0.032	44	0.032											BV
Crickhowell	SO207202	24	0.03	21	0.03	27	0.03											AV
Crucorney	SO323221	24	0.0022	27	0.0022	21	0.0022											AV
Crumlin	ST228984	60	0.002	57	0.002	53	0.002											CDV
Cwm Ffrwd-oer	SO265014	43	0.002	46	0.002	50	0.002											BV
Cwmafon	SS798936	24	0.014	21	0.014	27	0.014											AV
Cwmaman	ST000993	49	0.002	42	0.002	45	0.002											BV
Cwmfelinfach	ST184909	48	0.004	52	0.004	56	0.004											CDV
Deri	SO121022	25	0.001	22	0.001	28	0.001											AV
Dowlais	SO073088	49	0.01	58	0.01	54	0.01											CDV
Ebbw Vale	SO159088	59	0.1	55	0.1	50	0.1											CDV
Ebbw Vale South	SO176073	24	0.004	27	0.004	21	0.004											AV
Ferndale	ST006970	60	0.02	57	0.02	53	0.02											CDV
Fernhill	ST030993	59	0.002	55	0.002	50	0.002											CDV
Gelli-fendigaid	ST070935	59	0.0024	55	0.0024	50	0.0024											CDH
Gilfach	SS982890	24	0.01	21	0.01	27	0.01											AV
Llanfach	ST217947	60	0.002	57	0.002	53	0.002											CDH
Llanfoist	SO308143	60	0.014	53	0.014	57	0.014											CDV
Llangeinor	SS905886	59	0.038	55	0.038	50	0.038											CDV
Llanharan	SS998831	24	0.002	21	0.002	27	0.002											AV
Llanhilleth	SO213004	39	0.01	42	0.01	45	0.01											BV
Llyswen	SO137361	24	0.006	27	0.006	21	0.006											AV
Machen Upper	ST211897	50	0.007	55	0.007	59	0.007											CDV
Maesteg	SS841913	25	0.05	22	0.05	28	0.05											AV
Merthyr Tydfil	SO057065	25	0.025	22	0.025	28	0.025											AV
Monmouth	SO526128	59	0.046	55	0.046	50	0.046											CDV
Mynydd Bach	ST168925	49	0.05	58	0.05	54	0.05											CDV
Mynydd Machen	ST223900	23	0.4	26	0.4	29	0.4											AV
Nant-y-Moel	SS934935	27	0.002	21	0.002	24	0.002											AV
Nantyglo	SO189106	60	0.002	57	0.002	53	0.002											CDH
Ogmore Vale	SS929894	60	0.02	57	0.02	53	0.02											CDV

Television Viewer's Guide

FREEVIEW DIGITAL TRANSMITTER SITES

Transmitter	NGR	PSB1 BBC A		PSB2 D3&4		PSB3 BBC B		COM4 SDN		COM5 Arqiva A		COM6 Arqiva B		COM7 Arqiva C		COM8 Arqiva D		Pol
		Ch	kW	Ch	kW	Ch	kW	Ch	kW	Ch	kW	Ch	kW	Ch	kW	Ch	kW	
Pennar	ST209958	43	0.02	46	0.02	40	0.02											BV
Pennorth	SO103266	23	0.01	29	0.01	26	0.01											AV
Penrhiwceiber	ST066966	53	0.002	57	0.002	60	0.002											CDV
Pontypridd	ST085905	25	0.4	22	0.4	28	0.4											AV
Porth	ST029919	43	0.012	46	0.012	40	0.012											BV
Rhondda Fach	ST007939	25	0.002	22	0.002	28	0.002											AV
Rhondda	SS990938	23	1	26	1	29	1											AHV
Rhymney	SO127042	60	0.03	57	0.03	53	0.03											CDV
Risca	ST240905	43	0.002	46	0.002	40	0.002											BV
Sennybridge	SN914295	43	0.013	50	0.013	46	0.013											BV
South Maesteg	SS860897	59	0.002	55	0.002	50	0.002											CDV
South Tredegar	SO155060	39	0.0026	42	0.0026	45	0.0026											BV
Taffs Well	ST123848	59	0.01	55	0.01	50	0.01											CDV
Ton Pentre	SS960955	49	0.032	58	0.032	54	0.032											CDV
Tonypandy	SS986924	59	0.002	55	0.002	50	0.002											CDV
Tonyrefail	ST009874	59	0.004	55	0.004	50	0.004											CDV
Trebanog	ST020907	24	0.002	27	0.002	21	0.002											AV
Trecastle	SN885274	25	0.002	22	0.002	28	0.002											AV
Trefechan	SO030085	42	0.01	39	0.01	45	0.01											BV
Treharris	ST103964	52	0.01	48	0.01	56	0.01											CDV
Tynewydd	SS931993	59	0.004	50	0.004	55	0.004											CDV
Upper Killay	SS590927	24	0.002	21	0.002	27	0.002											AV
Usk	SO384006	48	0.0036	52	0.0036	56	0.0036											CDH
Van Terrace	ST168865	39	0.002	45	0.002	42	0.002											BV
Wattsville	ST215911	60	0.002	57	0.002	53	0.002											CDV
Ynys Owen	ST082992	59	0.016	55	0.016	50	0.016											CDV
Whitehawk Hill	TQ330045	60	4	53	4	51	4	57	4	56	4	48	4					CDV
Bevendean	TQ334066	43	0.005	40	0.005	46	0.005											WV
Brighton Central	TQ312049	41	0.01	47	0.01	44	0.01											BH
Coldean	TQ329083	44	0.008	47	0.008	41	0.008											BV
Hangleton	TQ269077	50	0.005	55	0.005	52	0.005											CDV
Ovingdean	TQ358044	49	0.01	50	0.01	55	0.01											CDVH
Patcham	TQ318093	43	0.014	50	0.014	46	0.014											BH
Portslade	TQ245080	41	0.01	47	0.01	44	0.01											BV
Saltdean	TQ388034	55	0.02	54	0.02	58	0.02											CDV
Winter Hill	SD660144	50	100	59	100	54	100	58	100	49	100	55	100	31	25	37	23	WH
Winter Hill (Gtr Man local mux)		57	1															
Austwick	SD783672	45	0.006	39	0.006	42	0.006											BV
Backbarrow	SD358841	53	0.002	60	0.002	57	0.002											CDV
Bacup	SD878224	46	0.05	43	0.05	40	0.05											BV

Television Viewer's Guide

FREEVIEW DIGITAL TRANSMITTER SITES | 141

Transmitter	NGR	PSB1 BBC A		PSB2 D3&4		PSB3 BBC B		COM4 SDN		COM5 Arqiva A		COM6 Arqiva B		COM7 Arqiva C		COM8 Arqiva D		Pol
		Ch	kW	Ch	kW	Ch	kW	Ch	kW	Ch	kW	Ch	kW	Ch	kW	Ch	kW	
Barrow Town Hall	SD198691	44	0.002	41	0.002	47	0.002											BV
Birch Vale	SK028861	46	0.05	43	0.05	40	0.05											BV
Blackburn	SD703276	46	0.02	43	0.02	40	0.02											BV
Bollington	SJ941778	27	0.004	24	0.004	21	0.004											AV
Brinscall	SD630223	21	0.002	24	0.002	27	0.002											AV
Broadbottom	SJ987933	45	0.002	42	0.002	39	0.002											BV
Brook Bottom	SD969029	53	0.002	60	0.002	57	0.002											CDV
Burbage	SK044726	47	0.003	41	0.003	44	0.003											BV
Buxton	SK060753	27	0.2	24	0.2	21	0.2											AV
Cartmel	SD375793	28	0.002	25	0.002	22	0.002											AH
Chaigley	SD686446	27	0.002	24	0.002	21	0.002											AV
Chatburn	SD765445	26	0.002	23	0.002	29	0.002											AV
Chinley	SK035827	53	0.002	60	0.002	57	0.002											CDV
Congleton	SJ865619	44	0.04	41	0.04	47	0.04											BV
Dalton	SD230745	46	0.005	43	0.005	40	0.005											BV
Darwen	SD708223	45	0.1	49	0.1	42	0.1											BV
Delph	SD987080	26	0.002	23	0.002	29	0.002											AV
Dog Hill	SD951091	46	0.017	43	0.017	40	0.017											BV
Elton	SJ457735	27	0.013	24	0.013	21	0.013											AV
Far Highfield	SD543672	52	0.003	56	0.003	48	0.003											CDH
Glossop	SK027953	28	0.05	25	0.05	22	0.05											AV
Haslingden	SD795236	26	2	23	2	29	2											AV
Haughton Green	SJ934925	46	0.002	43	0.002	40	0.002											BH
Ladder Hill	SK027789	26	0.2	23	0.2	29	0.2											AV
Langley	SJ938709	27	0.002	24	0.002	21	0.002											AV
Lees	SD961039	28	0.002	25	0.002	22	0.002											AH
Littleborough	SD950166	27	0.1	24	0.1	21	0.1											AV
Macclesfield	SJ925725	28	0.007	25	0.007	22	0.007											AV
Manchester Hulme	SJ828966	44	0.002	41	0.002	47	0.002											BV
Melling	SD602703	57	0.005	60	0.005	53	0.005											CDHV
Middleton	SD876058	28	0.008	25	0.008	22	0.008											AV
Millom Park	SD162830	28	0.043	25	0.043	22	0.043											AV
Moss Bank	SJ509974	27	0.002	24	0.002	21	0.002											AV
Mottram	SJ988961	46	0.002	43	0.002	40	0.002											BV
Newchurch	SD840226	21	0.002	24	0.002	27	0.002											AH
Norden	SD861142	53	0.002	57	0.002	60	0.002											CDV
North Oldham	SD928059	27	0.008	24	0.008	21	0.008											AV
Oakenhead	SD806234	44	0.02	41	0.02	47	0.02											BV
Over Biddulph	SJ896605	57	0.004	53	0.004	60	0.004											CDV
Parbold	SD477112	44	0.03	41	0.03	47	0.03											BV
Penny Bridge	SD311836	26	0.006	23	0.006	29	0.006											AV
Portwood	SJ908911	28	0.002	25	0.002	22	0.002											AV

Television Viewer's Guide

142 FREEVIEW DIGITAL TRANSMITTER SITES

Transmitter	NGR	PSB1 BBC A		PSB2 D3&4		PSB3 BBC B		COM4 SDN		COM5 Arqiva A		COM6 Arqiva B		COM7 Arqiva C		COM8 Arqiva D		Pol
		Ch	kW	Ch	kW	Ch	kW	Ch	kW	Ch	kW	Ch	kW	Ch	kW	Ch	kW	
Prestbury	SJ885767	46	0.002	43	0.002	40	0.002											BV
Ramsbottom	SD803159	53	0.016	60	0.016	57	0.016											CDV
Ribblesdale	SD814719	44	0.006	41	0.006	47	0.006											BV
Romiley	SJ954904	44	0.002	41	0.002	47	0.002											BV
Roose	SD220692	22	0.02	28	0.02	25	0.02											AHVD
Skelmersdale	SD502074	46	0.1	43	0.1	40	0.1											BH
Staveley-in-Cartmel	SD383854	46	0.002	43	0.002	40	0.002											BV
Stockport	SJ867904	24	0.002	21	0.002	27	0.002											AH
Sunningdale	SJ270896	44	0.03	41	0.03	47	0.03											BV
Trawden	SD909378	53	0.04	60	0.04	57	0.04											CDV
Urswick	SD263739	44	0.002	41	0.002	47	0.002											BV
Walton-le-Dale	SD545291	27	0.002	24	0.002	21	0.002											AH
Wardle	SD915172	28	0.002	25	0.002	22	0.002											AH
West Kirby	SJ224862	27	0.015	24	0.015	21	0.015											AV
Whaley Bridge	SK010815	45	0.002	39	0.002	42	0.002											BV
Whalley	SD729352	46	0.01	43	0.01	40	0.01											BV
Whitewell	SD833245	53	0.016	60	0.016	57	0.016											CDV
Whitworth	SD886203	28	0.05	25	0.05	22	0.05											AV
Woodnook	SD762277	45	0.002	39	0.002	42	0.002											BV

New local TV multiplexes

A government initiative, licensed by Ofcom, will lead to a network of local television stations coming on air during the next couple of years. The first station, Estuary TV, serving Grimsby and part of East Yorkshire, launched in November 2013. We have listed details of the stations already licensed, with their names and on-air dates where known.

You'll find a map with the locations of the initial wave of local TV stations and further information about local TV on the following pages. You can keep up with local TV's progress at the following websites.
www.localtv.org.uk
www.localtvmux.com.

Area	Channel name	Transmitter	Transmiotter grid reference	Ch/Pol	ERP:kW	In existing group aerial	Target launch
Aberdeen	Around Aberdeen	Durris	NO763899	30H	10	Yes	2015/16
Ayr	Ayrshire Today	Darvel	NS557341	30H	1	Yes	2015/16
Basingstoke	That's Hampshire	Hannington	SU527568	29H	2	No	2015/16
Belfast	NVTV	Divis	IJ287750	30H	5	Yes	On Air
Birmingham	Big Centre TV	Brierley Hill	SO916856	29V	0.2	No	On Air
	Big Centre TV	Sutton Coldfield	SK113003	51H	10	Yes	On Air
Brighton	Latest TV	Whitehawk Hill	TQ330045	54V	0.4	Yes	On Air

FREEVIEW LOCAL TV MULTIPLEXES 143

Area	Channel name	Transmitter	Transmiotter grid reference	Ch/Pol	ERP:kW	In existing group aerial	Target launch
Bristol	Made in Bristol	Bristol Ilchester Cres	ST577700	30V SFN	0.02	No	On Air
	Made in Bristol	Bristol Kings Weston	ST547775	30V SFN	0.02	No	On Air
	Made in Bristol	Mendip	ST563488	51H	10	Yes	On Air
Cambridge	Cambridge Presents	Madingley	TL329045	40H	1	Yes	On Air
Cardiff	Made in Cardiff	Wenvoe	ST110742	51H	10	Yes	On Air
Carlisle	That's Carlisle	Caldbeck	NY299425	56H	5	No	2015/16
Dundee	View from Bridges	Angus	NO394407	48H	1	Yes	2015/16
	View from Bridges	Tay Bridge	NO430284	51V	0.1	Yes	2015/16
Edinburgh	STV Edinburgh	Craigkelly	NT233872	30H	5	Yes	On Air
Glasgow	STV Glasgow	Black Hill	NS828647	51H	5	Yes	On Air
Grimsby	Estuary TV	Belmont	TF218836	27H	5	Yes	On Air
Guildford	That's Surrey	Guildford	SU975487	51V	0.1	Yes	2015/16
Leeds	Made in Leeds	Beecroft Hill	SE237350	56V SFN	0.02	Yes	On Air
	Made in Leeds	Emley Moor	SE222128	56H SFN	5	No	On Air
Liverpool	Bay TV	Storeton	SJ314841	30V	0.06	Yes	On Air
	Bay TV	Winter Hill	SD660144	56H	2	Yes	On Air
London	London Live	Crystal Palace	TQ339712	29H SFN	20	Yes	On Air
	London Live	Croydon	TQ332696	29H SFN	4	Yes	On Air
Maidstone	KMTV	Bluebell Hill	TQ757614	27H	1	Yes	2015/16
Manchester	That's Manchester	Winter Hill	SD660144	56H	2	Yes	On Air
Middlesbrough	Made in Teesside	Bilsdale	SE553962	24H	5	Yes	2015/16
Mold	Bay TV Clwyd	Moel-y-Parc	SJ123701	56H	2	Yes	2015/16
Newcastle	Made in Tyne & Wear	Pontop Pike	NZ148526	56H	5	Yes	On Air
Norwich	Mustard	Tacolneston	TM131959	57H	10	Yes	On Air
Nottingham	Notts TV	Nottingham	SK503435	50V	0.1	Yes	On Air
	Notts TV	Waltham	SK809233	26H	5	Yes	On Air
Oxford	That's Oxford	Oxford	SP567105	29H	4	Yes	On Air
Preston	That's Lancashire	Winter Hill	SD660144	56H	1	Yes	On Air
Reading	That's Reading	Hannington	SU527568	TBA			2015/16
Sailsbury	That's Sailsbury	Sailsbury	SU136285	51V	0.2	Yes	2015/16
Scarborough	Yorkshire Coast TV	Oliver's Mount	TA040869	56V	0.2	Yes	2015/16
Sheffield	SLTV Sheffield Live	Sheffield	SK324870	55V	0.1	Yes	On Air
Southampton	That's Solent	Rowridge	SZ447865	29H	10	Yes	On Air
Swansea	Bay TV Swansea	Kilvey Hill	SS671940	30V	0.1	Yes	2015/16
York	The York Channel	Bilsdale	SE553962	24H	2	Yes	2015/16

SFN - Two or more transmitters operating on the same frequency in a non-interfering manner.

Television Viewer's Guide

Local television stations

By the end of 2015, 28 local TV stations were on air. Others will follow in 2016 and 2017.

If it hasn't already arrived, local TV may be coming soon to a screen near you. The government initiative, licensed by Ofcom, will lead to more local television stations coming on air during the next couple of years. The first station, Estuary TV, serving Grimsby and part of East Yorkshire, launched in November 2013.

Local TV was initially funded by BBC licence money and then sustained by advertising and sponsorship. The infrastructure is provided by Comux Ltd, a community-owned organisation, while the stations themselves are provided and run by commercial operators who bid Ofcom for their licences, typically valid for up to 12 years.

To get local TV off the ground the BBC is contributing £25m towards capital costs and up to £5m per year over three years as a contribution to running expenses.

Locations

In the first phase of local TV, licences were granted for the following areas: Belfast, Birmingham, Brighton and Hove, Bristol, Cardiff, Edinburgh, Glasgow, Grimsby, London, Leeds, Liverpool, Manchester, Newcastle, Norwich, Nottingham, Oxford, Preston, Sheffield and Southampton.

Transmitters

Local TV is broadcast from existing Freeview transmitter sites. It uses main and relay masts with low power emissions.

How to get local TV

For Freeview viewers no extra aerial or receiver will be needed, but you may need to retune. Thereafter you need only select channel 8 in the UK and Northern Ireland, or channel 34 in Wales and Scotland. Sky has several local stations on its channel 117, and Virgin on 159. It is likely that operators will offer their programmes on Freesat and online.

Programmes and contributors

The ethos of local TV is to give communities access to television. Local stories can be dealt with in more depth than is afforded by BBC and ITV regional stations.

Some local stations, such as London Live, provide 24-hour coverage. Others run for a limited number of hours during the day depending on demand, advertising and the number of people served.

Programmes include local news and weather, travel, phone-ins, current affairs as they impinge on the local populace, home and property programmes and sports. Other programmes address community affairs and feature cultural, arts and factual subjects, in some ways like the coverage presently given by BBC regional radio programmes.

A range of people contribute to local TV, chosen for the relevance and interest of their input; substantial contributors may be the local newspaper (who may own the station) town councillors, MPs and other prominent figures.

Progress with local TV

Following the launch of Estuary TV in late 2013 the next year saw another seven stations begin broadcasting. For some of them there has been a bumpy ride with, for instance, London Live applying to Ofcom to reduce its operating hours; in late 2015 its whole future was under review. Other stations have not lived up to their promoters' hopes, while Birmingham's Centre TV changed hands before launch. Not all local TV stations will be successful in the face of competition from other media.

Further information

See page 142 for on-air dates and local TV channel service names.
The following websites have useful info.
www.localtv.org.uk
www.terrestrialtv.co.uk

LOCAL TV STATIONS | 145

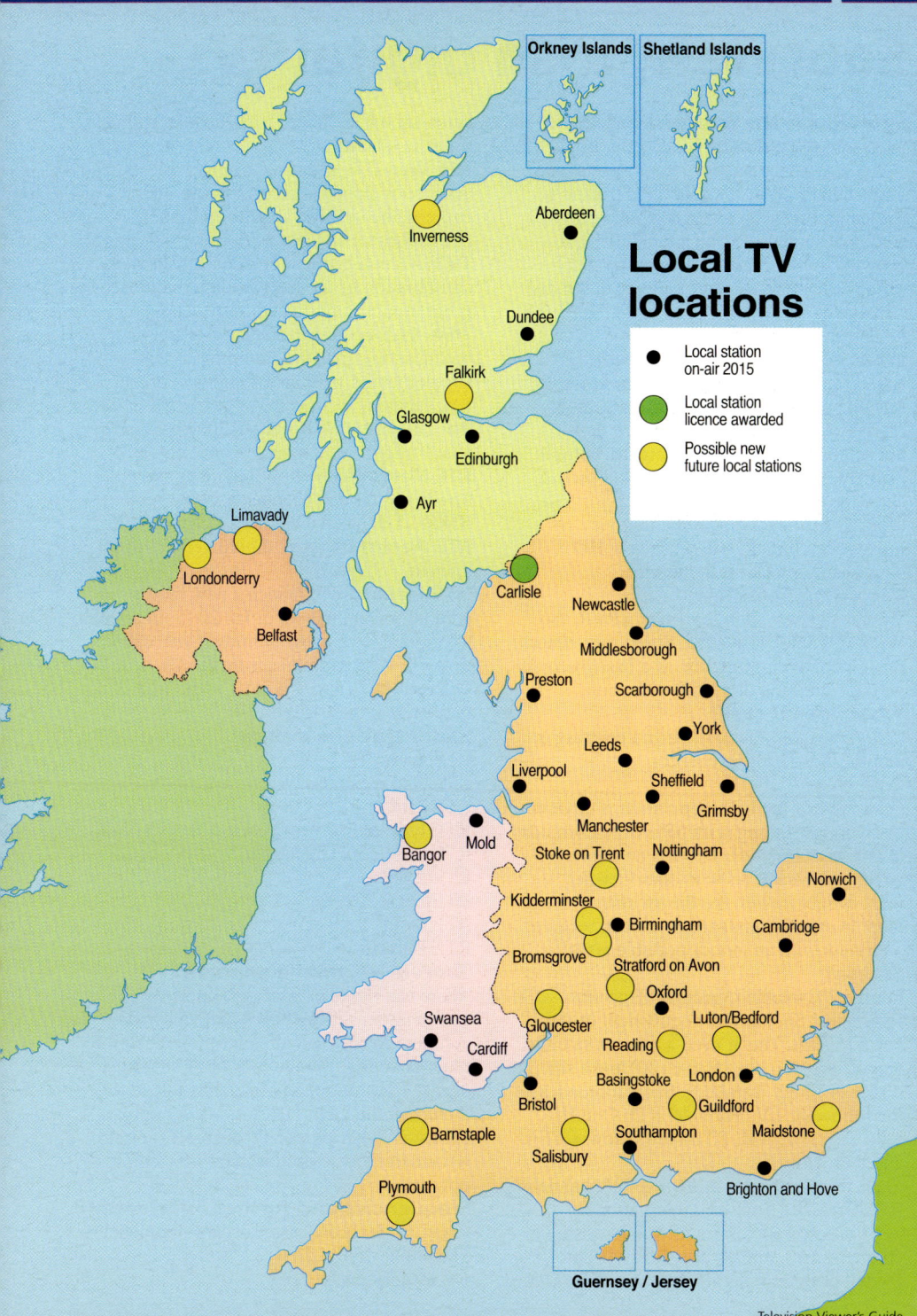

Saorview, Irish digital TV

Saorview is the national digital terrestrial television service for the Republic of Ireland. The Irish word Saor means free. Launched in 2011, the service is Ireland's source of terrestrial television following analogue switchover in October 2012. It is owned by Irish network operator 2RN (RTÉNL) whose carriage charges dissuaded Oireachtas TV and TV3 HD from launching in 2015.

What equipment do you need?
Like Freeview, Saorview requires an aerial, set-top box (€40 to €80) or an integrated digital television(€230). Saorview does not sell or produce any equipment, but it is available from official retailers and should carry the official Saorview logo. Saorview uses MPEG4 encoding, enabling viewers in Northern Ireland with Freeview HD equipment to pick up the service.

Recording Saorview
Walker (www.walker.ie) offers a PVR, the WP6500TTR, €110-€219. Some twin-tuner set-top boxes can function as a basic recorder with the addition of a USB disk drive.

What does it cost?
Saorview, like Freeview, has no contract or monthly subscription charges.

Coverage
Over 98% of the Irish population is covered by Saorview (along with 65% of households in Northern Ireland). There is an online coverage checker at: www.saorview.ie/what-is-saorview/make-the-switch/coverage-map In strong signal areas you can receive Saorview with an indoor portable aerial.

Saorsat
For the 2% unable to receive Saorview RTÉ offers Saorsat - the equivalent of the UK's Freesat. You'll need a Saorsat satellite receiver set-top box (€70-€120) and a satellite dish(€240 fully installed). Saorsat broadcasts the following TV channels: RTÉ One HD, RTÉ Two HD,TG4, RTÉ News Now, RTÉjr, and RTÉ One+1.
Saorview carries the following TV channels:
RTÉ One HD Irish and international news and current affairs as well as Irish entertainment, drama, arts, lifestyle and factual programming. RTÉ One+1 SD offers a catch-up for the RTÉ One schedule, showing it one hour later.
RTÉ Two HD offers a range of Irish sport, and young people's programming. In the evening the channel broadcasts drama, entertainment and acquisitions.
TV3 An independent broadcaster shows a combination of the Irish and international programmes, news and current affairs.
TG4 Irish language television channel, broadcasting Irish programming supplemented with programmes in other languages.
3e This channel broadcasts dramas, comedy, sports and movies mainly from USA/UK.
RTÉ News Now A rolling news and on-screen digital text channel.
RTÉjr A dedicated children's channel.
RTÉ Aertel Digital digital version of RTÉ teletext.
UTV Ireland will launch a new family entertainment channel in Jan 2015. In addition to original Irish programming it will carry many ITV programmes from the UK.
Oireachtas (Parliament) TV and Irish TV may launch in 2015 as well.
Radio stations include:
RTÉ Radio 1
RTÉ Radio 1 Extra
RTÉ 2fm
RTÉ lyric fm
RTÉ Raidió na Gaeltachta
RTÉ 2XM
RTÉjr/RTE Chill
RTÉ Gold
RTÉ Pulse

Saorview Combi equipment
There is a range of Saorview approved combination set-top boxes and TVs that combine Saorview with a free-to-air satellite tuner enabling users to pick up Irish terrestrial digital TV and free satellite channels, including the main UK stations. A popular choice is the Walker WP645TS-HD at €89. www.saorview.ie/product/saorview-combi-options/
Frequencies and further information
www.2rn.ie/ broadcast/saorview-frequencies/
Telephone 1890 222 012
www.saorview.ie/

CONTACTS AND FURTHER INFORMATION

BBC NATIONAL CONTACTS

BBC Broadcasting House
Portland Place, London
W1A 1AA
tel 0370 901 1227

BBC Media Village
201 Wood Lane, London
W12 7TQ
tel 020 8743 8000

BBC Television Centre
Wood Lane, London
W12 7RJ

BBC main switchboard
tel 020 7580 4468
www.bbc.co.uk

BBC Salford
MediaCityUK, Bridge
House, Salford Quays,
Manchester M50 2BH
tel 0161 836 0010
mediacityuk.co.uk

BBC Information & Reception Advice
PO Box 1922, Darlington
DL3 0UR tel 03700 100 123
reception@bbc.co.uk
www.bbc.co.uk/reception

Radio and TV Investigation Service
has been set up to help viewers and listeners investigate reception and interference problems, based at the BBC. PO Box 1922, Darlington DL3 0UR
tel 03709 016 789
www.radioandtvhelp.co.uk

BBC Press Office
A source of press releases from the BBC.
www.bbc.co.uk/mediacentre

BBC REGIONS

BBC English Regions HQ is based at BBC Birmingham
www.bbc.co.uk/england/tv/index.shtml

London
BBC London News,
Egton Wing, Portland Place,
London W1A 1AA
tel 020 7765 2667
yourlondon@bbc.co.uk

South East
South East Today,
The Great Hall, Mount
Pleasant Road, Tunbridge
Wells, Kent TN1 1QQ
tel 01892 675580
southeasttoday@bbc.co.uk

South
South Today, Broadcasting
House, Havelock Road,
Southampton SO14 7PU
tel 023 8022 6201
south.today@bbc.co.uk

Channel Islands
BBC Channel Islands
News,18 Parade Road,
St Helier JE2 3PL
tel 01534 837260
cinews@bbc.co.uk

South West
BBC South West,
Broadcasting House,
Seymour Road, Plymouth
PL3 5BD tel 01752 229 201
spotlight@bbc.co.uk

West
BBC Points West,
Broadcasting House,
Whiteladies Road, Bristol
BS8 2LR tel 0117 973 2211
pointswest@bbc.co.uk

Oxford
BBC Oxford,
269 Banbury Road,
Oxford OX2 7DW
tel 01865 311 444
oxford@bbc.co.uk

West Midlands
BBC West Midlands,
Level 7, The Mailbox,
Birmingham B1 1RF
tel 0121 567 6000
midlandstoday@bbc.co.uk

North West
BBC North West Tonight,
Quay House, Salford Quays,
Salford M50 2QH
tel 0161 335 6820
nwt@bbc.co.uk

North East and Cumbria
BBC Look North,
Broadcasting Centre, Barrack
Road, Newcastle NE99 2NE
tel 0191 232 1313
look.north.comment@bbc.co.uk

Yorkshire
BBC Look North,
2 St Peters Square,
Leeds LS9 8AH
tel 0113 224 7041
look.north@bbc.co.uk

East Yorks and Lincs
Look North,
Queen's Court, Queen's
Gardens, Hull HU1 3RH
tel 01482 323232
looknorth@bbc.co.uk

East Midlands
East Midlands Today,
London Road, Nottingham
NG2 4UU tel 0115 902 1930
emt@bbc.co.uk

East
BBC Look East,
The Forum, Millennium
Plain, Norwich NR2 1BH
tel 03457 630630
look.east@bbc.co.uk

Cambridge
BBC Look East
Cambridge Business Park,
Cowley Road, Cambridge
CB4 0WZ
tel 01223 259 696
look.east@bbc.co.uk

BBC Northern Ireland
25 Ormeau Avenue, Belfast
BT2 8HQ tel 028 9033 8000
www.bbc.co.uk/northernireland

BBC Scotland
40 Pacific Quay, Glasgow
G51 1DA
tel 0141 422 6000
www.bbc.co.uk/scotland

BBC Wales
Broadcasting House,
Llandrisant Road, Cardiff
CF5 2YQ
tel 029 2032 2000
feedback@wales.bbc.co.uk
www.bbc.co.uk/wales

CONTACTS AND FURTHER INFORMATION | 149

BBC CHANNELS

BBC main channels have a website with useful information and links to their main programmes. They can be accessed from www.bbc.co.uk/tv Click on the channel logo to open the channel's website.

ITV (CHANNEL 3)

ITV is not owned by a single company. Instead four companies currently provide the regional services.

Anglia TV
Anglia House, Norwich
NR1 3JG tel 0844 8816 920
www.itv.com/news/anglia

London ITV
200 Gray's Inn Road,
London WC1X 8XZ
tel 020 7430 4000
www.itv.com/news/london

ITV Central
Central Court, Gas Street,
Birmingham B1 2JT
tel 08448 814 000
www.itv.com/news/central

Channel Television
Le Capelain House, Castle Quay, St Helier, Jersey JE2 3EH tel 01534 480 526
Television House, Bulwer Avenue, St Sampson, Guernsey GY2 4LA tel 01481 241 882
www.itv.com/news/channel//

STV North/ STV Central
Pacific Quay, Glasgow
G51 1PQ   tel 0141 300 3000
www.stv.tv

Granada ITV
Orange Tower, Media City,
Salford M50 2HE
tel 0161 9526073
www.itv.com/news/granada

Wales ITV
Media Centre,
Culverhouse Cross, Cardiff
CF5 6XJ tel 0844 881 0100
www.itv.com/news/wales/

Meridian ITV
Fusion 3, 1200 Parkway,
Fareham PO15 7AD
tel 0844 881 2000
www.itv.com/news/meridian

Tyne Tees & Border ITV
ITV House, The Watermark, Gateshead, Tyne and Wear NE11 9SZ
tel 0844 881 5000
www.itv.com/news/tyne-tees
www.itv.com/news/border

Ulster Television
Ormeau Road, Belfast BT7 1EB tel 028 9032 8122
www.u.tv

Westcountry ITV
Bath Road, Bristol
BS4 3HG tel 0844 881 2307
www.itv.com/news/westcountry

Yorkshire ITV
TV House,106 Kirkstall Road, Leeds LS3 1JS
tel 0113 222 7200
www.itv.com/news/calendar

ITV Network Centre
200 Gray's Inn Road,
London WC1X 8HF
tel 020 7156 6000
www.itv.com

ITV Studios UK
The London Television Centre, Upper Ground
London SE1 9LT
tel 020 7491 1441

ITV Viewers Services
Gas Street, Birmingham B1 2JT tel 0344 88 14150
viewersservices@itv.com

Good Morning Britain
The London Television Centre, Upper Ground,
London SE1 9LT
tel 0344 88 14150 (option1)
www.itv.com/goodmorningbritain

Independent Television News Ltd (ITN)
200 Gray's Inn Road,
London WC1X 8XZ
tel 0207 430 4622
www.itn.co.uk

CHANNEL 4 / FIVE

Channel 4
124-126 Horseferry Road,
London SW1P 2TX
tel 0207 396 4444
www.channel4.com

Channel Five
10 Lower Thames St,
London EC3R 6EN
tel 020 8612 7700 or 03457 050505
www.channel5.com

S4C
Parc Ty Glas, Llanishen
Cardiff CF14 5DU
tel 0370 600 4141
www.s4c.co.uk/cy

FREEVIEW CHANNELS

Al Jazeera
1 Knightbridge, London
SW1X 7XW
020 7201 2800
www.aljazeera.com

Capital TV
30 Leicester Square
London WC2H 7LA
020 7054 8000
www.capitalfm.com/tv

CBBC and Cbeebies
Bridge House, Media City,
Salford M50 2BH
www.bbc.co.uk/cbbc
www.bbc.co.uk/cbeebies

Community Channel
4th Floor, Block A,
Centre House, Wood Lane
London, W12 7SB
020 7871 5600
www.communitychannel.org

Create and Craft / Ideal World Ideal Home House, Newark Road, Peterborough
PE1 5WG 08431 681 681
www.createandcraft.tv
www.idealworld.tv

Television Viewer's Guide

150 CONTACTS AND FURTHER INFORMATION

Dave, Drama, Really & Yesterday
10 Hammersmith Grove,
London W6 7AP
0203 752 7707
uktv.co.uk
Fire Hits
The Picture House, 307
Holdenhurst Road
Bournemouth BH8 8BX
01202 835 107
www.fireradio.co.uk
Food Network
111 Buckingham Palace
Road, London SW1W 0SR
020 578 9700
www.foodnetwork.co.uk
4 Music
Francis House, 11 Francis St,
London SW1P 1DE
020 7306 8087
www.4music.com
Gems TV
Eagle Road Studios,
Eagle Road, Redditch,
Worcestershire B98 9HF
tel 0333 400 0011
gems.tv
Jewellery Channel
PO Box 443, Feltham TW13
9DU tel 0344 375 2525
www.tjc.tv
Movies 4 Men and Sony SAB
25 Golden Square, London
W1F 9LU 020 7533 1000
www.sonytv.com
Peace TV
Quadrant Court, 48
Calthorpe Street, Edgbaston
Birmingham B15 1TH
0759 5381 379
www.peacetv.tv
Quest
Chiswick Park, Building 2
566 Chiswick High Road
London W4 5YB
020 8811 3419
www.questtv.co.uk
QVC
Chiswick Park, Building 8
566 Chiswick High Road
London W4 5XU

020 8811 5387
www.qvcuk.com
Racing Channel
3rd Floor
Gillingham House
38-44 Gillingham Street
London SW1V 1HU
020 7592 0457
www.racinguk.com
RT (Russia Today)
Building 1
4 Zubovsky Boulevard
Moscow 119021
www.rt.com
Travel Channel
11 Buckingham Palace Rd,
London SW1W 0SR
tel 0207 578 9700
www.travelchannel.co.uk
VIVA/ MTV
17-29 Hawley Crescent,
Camden, London NW1 8TT
tel 0203 580 2000
www.viva.tv or mtv.co.uk

OTHER BROADCASTERS

Amazon Fire TV
www.amazon.co.uk
Astra
3 Dorset Rise, London
EC4Y 8EN
tel 020 7632 7920
www.ses.com
BT TV
tel 0800 800 150
Customer support
tel 0800 111 4567
bttvproductsandservices.bt.com
BT Sport
London E20 3BS
sport.bt.com
EE TV
tel 0800 956 000
ee.co.uk
Freesat
Freesat Enquiries, 23-24
Newman Street, London
W1T 1PJ
tel 0345 313 0051
www.freesat.co.uk

Freeview
Freeview, DTV Services
Ltd, 27 Mortimer St, London
W1T 3JF
Freeview advice
tel 03456 505050
www.freeview.co.uk
Plusnet
tel 0800 432 0200
www.plus.net
Sky
Grant Way, Isleworth
TW7 5QD
tel 03442 411 653
www.sky.com
Freesat from Sky
tel 03442 410 595
skydigital@sky.com
Sky Technical assistance
Check Sky's website
www.sky.com/shop/tv/boxes/
Customer enquiries
tel 03442 411 411
TalkTalk TV
PO Box 360, Southampton
SO30 2LY
tel 0203 441 5550
Customer support tel
0870 444 1820
www.talktalk.co.uk/tv/
Virgin Media
10-14 Bartley Way, Hook,
Hampshire RG27 9UP
tel 0800 052 0626
www.virginmedia.com
YouView
10 Lower Thames Street
London EC3R 6YT
tel 020 3355 7955
www.youview.com

LOCAL TV STATIONS

Bay TV, Liverpool
0151 559 8688
www.baytvliverpool.com
Big Centre TV,
Birmingham 01922 438 106
www.bigcentre.tv
Cambridge Presents
tel 01223 750 890
www.cambridge-tv.co.uk

Television Viewer's Guide

CONTACTS AND FURTHER INFORMATION | 151

Estuary TV, Grimsby and **Yorkshire Coast TV,**
Scarborough 01472 315561
estuary.tv

Latest TV, Brighton
www.thelatest.tv

London Live
www.londonlive.co.uk

Made in Bristol
0117 9066 551
www.madeinbristol.tv

Made in Cardiff
07970 112 903
www.madeincardiff.tv

Made in Leeds
0113 200 7007
www.madeinleeds.tv

Made in Tyne and Wear,
Newcastle 07551 676 943
www.madeintyneandwear.tv

Maidstone KMTV,
tel 01634 227800
www.kentonline.co.uk

Mustard TV, Norwich
01603 628311
www.mustardtv.co.uk

Northern Vision Television NVTV, Belfast
028 9024 5495
www.northernvisions.org

Notts TV, Nottingham
0115 993 2350
nottstv.uk.com

Sheffield SLTV
0114 281 4082
web.sheffieldlive.org

STV Edinburgh & Glasgow
0141 300 3000 www.stv.tv

That's TV various locations
tel 020 7612 4141
thatstv.com

STATUTORY BODIES

Ofcom
Riverside House,
2a Southwark Bridge Road,
London SE1 9HA
tel 0300 123 3000 or
020 7981 3000
contact@ofcom.org.uk
www.ofcom.org.uk

Digital UK
The independent, non-profit organisation supporting Freeview viewers and channel providers.
tel 0345 650 50 50
www.digitaluk.co.uk

Department for Culture, Media and Sport
100 Parliament Street,
London SW1A 2BQ
tel 020 7211 6000
enquiries@culture.gov.uk
www.gov.uk

The European Broadcasting Union
L'Ancienne-Route 17A,
Postal Box 45, 1218 Le Grand-Saconnex, Geneva, Switzerland
tel + 41 (0)22 717 2111
ebu@ebu.ch www3.ebu.ch

TV Licensing
Darlington DL98 1TL
tel 0300 790 6112
www.tvlicensing.co.uk

INDUSTRY AND TRADE BODIES

BARB (Broadcasters' Audience Research Board) compiles television audience data in the UK. The company is jointly owned by the BBC, ITV, Channel 4, Channel 5, BSkyB and the IPA (Institute of Practitioners in Advertising). Limited channel and programme data is available from their website. Registration fees currently stand at £8500 together with quarterly fees that depend on the data required.
Switchboard 0207 024 8100
www.barb.co.uk

Confederation of Aerial Industries (CAI)
The CAI can help you if you want advice about aerials, dishes or installers.
Communications House, 41a Market Street, Watford. Herts WD18 0PN
tel 01923 803030
www.cai.org.uk

Digital Television Group
The DTG was formed in 1995 to set technical standards for the implementation of digital terrestrial television (DTT) in the UK. Its website has digital news, plus general and technical information about Freeview.
5th Floor, 89 Albert Embankment, London SE1 7TP
General tel 020 7840 6500
Consumer tel 0345 650 50 50
www.dtg.org.uk

IPA
Institute of Practitioners in Advertising
44 Belgrave Square
London SW1X 8QS
tel 020 7235 7020
www.ipa.co.uk

Retra (Radio, Electrical & Television Retailers' Association)
Retra House
St John's Terrace, 1 Ampthill Street, Bedford MK42 9EY
tel 01234 269110
retra@retra.co.uk
www.retra.co.uk

CONSUMER ORGANISATIONS

Which?
Customer Service Centre,
Castlemead, Gascoyne Way,
Hertford SG14 1LH
tel 01992 822800
www.which.co.uk

Voice of the Listener and Viewer (VLV)
A non-profit making society to support quality broadcasting.

Television Viewer's Guide

152 CONTACTS AND FURTHER INFORMATION

Members receive a quarterly newsletter. Membership £30 p.a. Concessions £22.50 p.a. The Old Rectory Business Centre, Springhead Road, North Fleet, Kent DA11 8HN
tel 01474 338 716
info@vlv.org.uk
www.vlv.org.uk

MANUFACTURERS

Amstrad tel 01277 228888
www.amstrad.com
Antiference
tel 01675 465487
www.antiference.co.uk
Archos tel 0208 822 3636
www.archos.com
Avtex tel 08448 806060
www.avtex.co.uk
Bang & Olufsen
tel 01189 692288
www.bang-olufsen.com
Bose tel 0808 1688 582
www.bose.co.uk
Bush Bought out by Argos
tel 0345 640 3030
technical support tel
0345 6040105
www.argos.co.uk
Echostar
tel 01535 659000
www.echostar-europe.com
Goodmans
tel 03444 122 541
www.goodmans.co.uk
Grundig UK
tel 0345 603 1234
www.grundig.co.uk
Hauppauge
tel 020 3405 1717
www.hauppauge.co.uk
Hitachi tel 0870 405 4405
www.hitachi.co.uk
Humax tel 0344 318 8800
www.humaxdigital.com/uk
Icecrypt was Topfield
tel 01795 429 666
www.icecrypt.com
JVC tel 0345 310 8000
www.jvc.co.uk
Kathrein
tel +49 8031 1840
No UK number
www.kathrein.co.uk
LG Electronics
tel 0344 847 5454
www.lg.com/uk
Loewe tel 0333 1230 220
or +49 9261 99500
www.loewe.tv/uk
Marantz tel 02890 279 830
www.marantz.co.uk
Maxview 01553 813300
www.maxview.co.uk
NEC tel 020 8836 2000
uk.nec.com
Netgem (France)
tel +33(0) 155 625562
www.netgem.com
One For All
tel 020 7744 0021
www.oneforall.co.uk
Pace tel 01274 532000
www.pace.com
Packard Bell
tel 0371 467 0008
www.packardbell.co.uk
Panasonic
tel 0844 844 3899
www.panasonic.co.uk
Philex Electronic Took over Labgear
tel 01234 263700
www.philex.com
Philips tel 020 7949 0241
www.philips.com
Sagem tel 01932 572 900
customer support tel
0845.0900 316
www.sagemcomdigital.co.uk
Samsung
tel 0330 726 7864
www.samsung.com/uk
Sharp tel 0800 262958
After sales 08705 274277
www.sharp.co.uk/gb
Slingbox
tel 1800 313 4274
uk.slingbox.com
Sony tel 0207 365 2413
www.sony.co.uk
Strong
(no UK number)
www.strongsat.tv
Tatung
www.tatung.com
Technisat
tel 0845 467 1935
www.technisat.com
Televes tel 01633 875821
www.televes.com
Thomson
The Strong Group has licensed the Thomson brand
Toshiba tel 01932 841 600
www.toshiba.co.uk
Triax tel 0845 601 0578
www.triax.co.uk
TVonics/ Pulse-Eight
Email enquiries
www.pulse-eight.com
Yamaha Electronics
tel 0844 811 1116
uk.yamaha.com

EQUIPMENT RETAILERS

Amazon tel 0844 545 6508
or 0800 496 1081
www.amazon.co.uk
Argos tel 0345 640 3030
www.argos.co.uk
Currys tel 03445 611234
www.currys.co.uk
John Lewis
tel 03456 049 049
www.johnlewis.com
Keene Electronics
tel 01332 830550
www.keene.co.uk
Lektropacks
tel 0844 209 1966
www.lektropacks.co.uk
Maplin Electronics
tel 0333 400 9500
www.maplin.co.uk
Richer Sounds
tel 0333 900 0094
www.richersounds.com
RoadPro tel 01327 312233
www.roadpro.co.uk
SatCure
www.satcure.co.uk
Sevenoaks Sound and Vision tel 01732 740 944
www.sevenoakssoundandvision.co.uk

Television Viewer's Guide

CONTACTS AND FURTHER INFORMATION | 153

Tesco Direct
tel 0800 323 4050
www.tesco.com/direct
Turbosat
tel 01795 429 666
www.turbosat.com

PROGRAMME GUIDES

Radio Times
Weekly, £2. Covers the main national networks with more detail than other guides. Covers digital TV channels, Sky Sports/Movies and has the best radio programme guide available.
Radio Times, Vineyard House, 44 Brook Green, London W6 7BT
tel 020 7150 5800
subs 0844 848 9729
www.radiotimes.com
TV Times
Weekly, £1.40. Has less detail about the main national networks than Radio Times, but covers a few more digital channels. Basic radio info. Published by Time Inc UK who also publish What's on TV and TV and Satellite Week. tel 020 3148 5000
subs 033 0333 1133
timeincuk.com
What's on TV
Weekly, 58p. Has less detail about the main national networks than TV Times with about the same number of digital channels covered. Basic radio coverage. Published by Time Inc UK
enquiries 020 3148 5000
subs 033 0333 1133
www.whatsontv.co.uk
TV Choice
Weekly, 50p. Regional programme guide with similar detail of main channels as What's on TV. No radio information. Published by Bauer.

Academic House, 24-28 Oval Rd, London NW1 7DT
tel 020 7241 8000
subs 01795 414831
www.tvchoicemagazine.co.uk
Total TV
Weekly, £1.10. Total TV Guide covers more than 90 channels every week, all in a clear, colour-coded and easy-to-follow format.
Published by Bauer.
www.bauer.co.uk
TV and Satellite Week
Weekly, £1.60. Digital & satellite channels covered. Useful weekly sports planner. Published by Time Inc UK
tel 020 3148 5000
subs 033 0333 1133
timeincuk.com
RTÉ Guide
Weekly TV and radio listings magazine for the Republic of Ireland. Some info available online. UK subs €148.72 p.a.
www.rteguide.rte.ie

ONLINE TV GUIDES

Most channels now offer a detailed programme schedule on their website. Other free services amalgamate programme information from a number of channels. Some offer programme alerts to remind you when a selected programme is due on air.
BBC TV programme information is available from www.bbc.co.uk/tv
Radio Times Online
A comprehensive, well designed and easy to use TV guide. www.radiotimes.com
TVGuide.co.uk
Comprehensive listings, with an easy-to-use design and offering a good degree of customisation.
The site carries information

for all the main UK TV platforms.
www.tvguide.co.uk
Sky Online Guide
Good range of channels and features, along with the ability to set programmes to record on a Sky+ box from the online listings.
tv.sky.com/tv-guide
ITV programme information available from www.itv.com/tvguide/
DigiGuide is a personalised TV guide linked to the internet. Digiguide. tvpremium is £2.99 p.a. A free iPhone app is available.
www.digiguide.tv
Euro TV website has ended. For French programmes try www.telepro.be and Dutch: www.tvgids.nl

TECHNICAL GUIDES

Newnes Guide to TV and Video Technology
4th edn., £28.99.
KF Ibrahim. Covers the fundamentals of digital TV and video. Order via Amazon. www.amazon.co.uk
Newnes TV and Video Engineer's Pocket Book
3rd edn., £34.99. Eugene Trundle. A guide to TV reception, satellite and cable TV, VCRs and fault finding. Order via Amazon.

CLUBS AND SOCIETIES

British FM and TV Circle
for TV and radio DXers.
skywaves.org
Royal Television Society
3 Dorset Rise, London EC4Y 8EN
tel 020 7822 2810
www.rts.org.uk

Television Viewer's Guide

154 CONTACTS AND FURTHER INFORMATION

COMPLAINTS

BBC
For general comments, queries and criticism
BBC Audience Services
PO Box 1922, Darlington
DL3 0UR.
tel 03700 100 222
www.bbc.co.uk/complaints

Ofcom
Ofcom handles complaints about programmes on ITV, Channel 4, Channel 5, satellite and cable. It has also replaced the Broadcasting Standards Commission (BSC) and now handles complaints about violence, sex and matters of taste and decency from any viewer or listener and on any radio or TV channel, and about unjust or unfair treatment or infringement of privacy from those directly affected by the programme.
Ofcom, Riverside House, 2a Southwark Bridge Road, London SE1 9HA
tel 0300 123 3333
consumers.ofcom.org.uk/complain

ITV Viewers Services
Gas Street, Birmingham B1 2JT tel 0344 88 14150
viewersservices@itv.com
itv.com/viewerservices

Channel 4
124-126 Horseferry Road, London SW1P 2TX
tel 0207 396 4444
www.channel4.com/4viewers

Channel Five
Customer Services
10 Lower Thames Street, London EC3R 6EN
tel 020 8612 7700
about.channel5.com/customer-services

INTERFERENCE PROBLEMS

The BBC has taken over the responsibility of investigating interference problems from Ofcom. Provided by the BBC, the Radio and Television Investigation Service helps viewers and listeners investigate and, where possible, resolve interference problems affecting their television and radio reception. The service covers interference to reception of all UK broadcasters. Check the link below to the Radio and Television Investigation Service.
www.radioandtvhelp.co.uk
Before they become involved you must have thoroughly investigated the problem.

Ofcom consumer advice
Ofcom also has simple guides on its website with information about a number of subjects including television.
consumers.ofcom.org.uk

at800 TV advice on 4G interference problems,
tel 0333 3131 800 or 0808 1313 800 (free from landlines and mobiles).
https://at800.tv

ENGINEERING WORKS

Digital UK is the organisation responsible for digital TV support in the UK. Its website lists planned engineering works.
www.digitaluk.co.uk

SIGHT OR HEARING PROBLEMS

BBC Subtitling service
Info about BBC subtitling.
Subtitling, BBC Television Centre, London W12 7RJ or Email subtitling@bbc.co.uk

Big Print is a weekly newspaper which is published with large bold type. It contains TV and radio listings. £100 for 12 months.
Big Print, RNIB 105 Judd St, London WC1H 9NE
tel 0303 123 999
helpline@rnib.org.uk
Online information about Big Print is available via the RNIB website.
www.rnib.org.uk

RNIB (Royal National Institute of the Blind) offers information, support and advice for people with sight problems.
RNIB, 105 Judd Street, London WC1H 9NE
tel 0303 123 9999
www.rnib.org.uk

Talking Newspaper Association provides programme listings on audio cassette and on computer disc. Based at RNIB 105 Judd St, London WC1H 9NE
tel 0303 123 9999
www.tnauk.org.uk

Action on Hearing Loss represents people who are deaf or hard of hearing in the UK (Formerly RNID). Its website has useful information and factsheet.
19-23 Featherstone Street, London EC1Y 8SL
tel 0808 808 0123
Textphone 0808 808 9000
information.line@hearingloss.org.uk
www.actiononhearingloss.org.uk

Sky's website has useful information and has widened the scope of its online TV listings, to give fuller weekly details of Sky

Television Viewer's Guide

CONTACTS AND FURTHER INFORMATION | 155

programmes which carry audio description, subtitles or sign language.
accessibility.sky.com
You can also call the Accessible Customer Service Team on 0344 241 0333
Textphone 0344 241 0535

GENERAL WEBSITES

Digital Spy The website is a good source for digital channel and technical news. It covers Sky, Freeview and cable with links to most UK channels.
www.digitalspy.co.uk

Media Guardian A good source of digital, television and media news stories.
www.theguardian.com/media

Advanced Television The site offers news about the television and broadcasting industry.
www.advanced-television.com

informitv Provides useful news and information about interactive, broadband, video-on-demand and personal television services.
http://informitv.com

Tech Watch Website with useful news about broadcast and associated technologies
www.techwatch.co.uk

TRANSMITTER / TECHNICAL WEBSITES

BBC Reception Advice Lots of useful information on digital and analogue television along with reception predictors and useful transmitter info. You'll also find details of transmitters that are off-air for maintenance or other reasons.
www.bbc.co.uk/reception

MB21
Mike Brown's website with lots of useful transmitter and television information.
tx.mb21.co.uk

Wolfbane Cybernetic's website has technical information and a postcode predictor for digital TV reception.
www.wolfbane.com

Wright's Aerials
Useful site with information about domestic and commercial aerial installation. Includes the 'Rogues Gallery' photos showing how not to install an aerial!
www.wrightsaerials.tv

DTG website
The Digital TV Group (DTG) is the industry association for digital television in the UK. The Group publishes and maintains the technical specification for the UK's Freeview and Freeview HD platforms.
www.dtg.org.uk

Antiference's website
The company manufactures TV aerial and satellite equipment. The website offers useful information with links to television and technical sites.
www.antiference.co.uk

megalithia.com
A useful site that enables you to plot your location and that of a transmitter to produce a line of sight diagram. The resulting plot shows your location and that of the transmitter, along with the distance and bearing to the transmitter.
www.megalithia.com/elect/terrain.html

INDUSTRY NEWS

Broadband TV News
Provides a regular summary of industry news in the TV industry, emailed to subscribers. Broadband TV News is available free to subscribers.
You can also subscribe to the daily email newsletter or the RSS news feed.
www.broadbandtvnews.com

Inside Satellite
Published every two weeks Inside Satellite covers what's happening in the DTH and communications satellite business. A one year subscription (24 issues) costs £497. PDF copies emailed.
insidesatellite.com

Broadcast
A weekly newspaper for the TV and Radio industries. TV news analysis and commentary plus weekly ratings data broken down by genre and channel. The company's website has useful information and an RSS news feed. £288 p.a.
tel 01604 82876
www.broadcastnow.co.uk

Digital TV Europe.net
is a monthly international magazine covering the business of broadband and pay-TV, whether cable, DTH, IPTV, DTT or mobile. It is published by Informa Telecoms and Media.
The website offers useful news and comment covering the TV industry. It offers a free daily newsletter.
www.digitaltveurope.net/register

Television Viewer's Guide

GLOSSARY ABBREVIATIONS

16:9 A widescreen picture format 16 units wide by 9 units high.

3D TV Three dimensional television - offers 3D images.

4:3 The traditional shape of TV screens before 16:9 (widescreen) came along.

4G 4G (4th Generation) the latest, and fastest, mobile phone standard. It may cause interference with Freeview terrestrial TV signals.

4K 4K resolution is a term for TV and other display devices, or content, having a horizontal resolution of approximately 4,000 pixels.

5.1 Sound system in use with DVD and home cinema systems using five main speakers and one subwoofer.

7.1 Sound system in use with Blu-ray and home cinema systems using seven main speakers and one subwoofer.

8K or Full Ultra HD (FUHD) is the highest ultra high definition television (UHDTV) resolution to exist in digital television and digital cinematography. 8K refers to the horizontal resolution of these formats, which are in the order of 8,000 pixels.

ADSL Asymmetric Digital Subscriber Line - a method of delivering television, video or a broadband internet connection using telephone wires - the asymmetric refers to the fact that download and upload speeds are not the same.

Analogue The traditional way in which TV pictures and sound have been transmitted or recorded. Not digital.

Anamorphic An image whose geometry is optimised for widescreen viewing. Viewed on a 4:3 TV it appears horizontally squashed.

App App is short for Application software. It is increasingly used in the mobile world where small applications can be downloaded to a smartphone or tablet device to increase functionality.

Artefacts Interference or distortion caused by a poor quality or overly compressed digital signal.

Aspect ratio The proportions or shape of a TV screen. For example 4:3, 16:9.

AD Audio Description. An additional soundtrack describing what's happening in the TV picture - very useful if you are blind.

AE Short for aerial. In practice it is used to label any plug or socket which carries an RF signal.

AV Audio Video. A general term used to describe the function of a socket, cable or other equipment.

AVI Designed by Microsoft, AVI (Audio Video Interleaved) is a widespread multipurpose file format used in computer video recordings.

Azimuth For satellite dish alignment - used to describe the horizontal angle between a fixed compass point (eg South) and a satellite position. For example Astra 2 satellites are at 28.2°East (of South).

BER Bit-Error Ratio.

Blocking See Artefacts.

Blue laser Laser technology used in high-definition technology such as Blu-ray and HD-DVD

Bluetooth A system enabling devices such as computers and printers to connect wirelessly.

Blu-ray A technology to increase the capacity of DVD discs.

BSC The Broadcasting Standards Commission.

C4 Channel 4.

C5 Channel Five.

CAI Confederation of Aerial Industries.

CAM Conditional Access Module. A viewing card fits into the CAM - necessary if you want to watch pay or subscription satellite channels.

CATV Cable TV.

CD Compact Disc. CD-R can be recorded on once, CD-RW can be recorded/erased many times.

Chipping Changing a computer chip inside a DVD player enabling it to play discs from round the world.

Chromecast a small WiFi media streaming device that plugs into the HDMI port on your TV.

CI Common Interface. A slot on a satellite receiver designed to take a CAM.

C/N Carrier-to-Noise ratio.

Co-ax Co-axial. Coaxial cable and connectors are commonly used to carry audio and video signals.

Component Video A way of sending video information between items of equipment.

Composite Video A relatively low quality way of sending video information between items of equipment.

CRT Cathode Ray Tube - traditional TV technology.

CSS Content Scrambling System - designed to prevent the unauthorised copying of standard-definition DVDs.

CVBS Colour, Video, Blank and Sync. An acronym often used to describe composite video.

D-CAB Digital Cable.

D-SAT Digital Satellite.

D-VHS Digital VHS, a high quality digital video signal recorded on tape - recordings can only be replayed in a D-VHS machine. Now obsolete.

DD Dolby Digital.

Digibox Another name for a digital set-top box.

DISEqC Digital Satellite Equipment Controller. Used in motorised satellite installations to control and position the satellite dish.

DivX A video codec created to compress lengthy video segments into small sizes while maintaining relatively high visual quality. It uses MPEG-4 compression. DivX has been controversial due to its use in the replication of copyrighted DVDs.

DLNA stands for Digital Living Network Alliance, and is basically a set of guidelines that electronics companies follow in order to enable devices to connect to each

GLOSSARY ABBREVIATIONS | 157

other. The technology makes it easier for consumers to use, share and enjoy their digital photos, music and videos.

DLP Digital Light Projection. A technology used in home cinema projectors and rear-pro TVs.

Dolby Dolby Surround/Pro-Logic/Digital - varieties of sound format.

DRM Digital Rights Management - technology designed to enable copyright holders to prevent unauthorised copying of their material.

DSO Digital Switch-off or Digital Switchover.

DSP Digital Signal Processing. Sound system that creates the illusion that you are listening in a hall, stadium, theatre etc.

DTG Digital Television Group.

DTH Direct to Home. Usually used to refer to Sky or other digital satellite services.

DTS Digital Theatre System. An alternative sound system to Dolby Digital. It uses five main speakers and a subwoofer.

DTT Digital Terrestrial TV. Includes Freeview.

DTV Digital Television.

DVB Digital Video Broadcast - sometimes qualified with a letter. For example DVB-S for satellite, DVB-S2 (more efficient version of DVB-S), DVB-T for terrestrial, DVB-H for handheld or mobile devices.

DVB-T2 A transmission standard that can be used to replace the exisiting DVB standard. It enables an increase in the channels that can be carried on Freeview by at least 30 percent and is used to carry HD TV signals.

DVD Digital Versatile Disc.

DVD-R DVD+R, DVD+RW, DVD-RW, DVD-RAM- different DVD recording formats.

DVI Digital Video Interface. A high quality video connection used by some PCs and high-definition TV screens. It does not carry audio signals.

DX Distance listening/viewing. Enthusiasts' reception of signals from distant transmitters.

EE TV New TV service launched in 2014 in the UK by telecommunications company EE. If offers a TV service, similar to that from BT TV and Talk Talk. It provides a mix of Freeview channels plus internet delivered content.

Encryption Encrypted or scrambled programmes require a subscription to be paid or a viewing card to be obtained before you can watch them.

EPG Electronic Programme Guide. An on-screen 'What's On' guide to television programmes.

ERP Effective radiated power.

Ethernet a type of network cabling used in computer networks – and in home systems often for connecting devices together or to a broadband router.

FAQs Frequently asked questions.

FF Fast Forward.

Fire TV Amazon's online TV platform. It gives access to a range of on-demand video content.

Firewire A high-speed coupling system commonly used with computers - also used to connect digital TV and video equipment. Also called i.LINK and IEE1394.

Footprint The area on the earth's surface covered by a satellite signal.

Freesat a BBC / ITV satellite service providing TV and radio station coverage for the UK.

FreesatfromSky Sky's package. This offers subscription free access to some of the TV and radio channels carried on Sky. For a one-off payment of £175 you receive a set-top box, satellite dish and installation.

Freeview The UK digital terrestrial television system launched in October 2002.

FTA Free-to-air - a TV or radio programme that is not encrypted and therefore does not require a subscription or a viewing card to receive.

FTV Free-to-view - a TV or radio programme that does not require subscription to receive but you may need to obtain a free viewing card.

GB or Gigabyte A unit used to measure the capacity of a disc, or computer hard disk drive.

Hacking A technique in which a code is entered via the remote control or a PC interface to change the region of a DVD player.

H.264 A standard for video compression. It is also known as MPEG-4 Part 10, or AVC.

HbbTV - stands for Hybrid Broadcast Broadband TV. What it means in practical terms is that you can get both standard broadcast free-to-air (FTA) TV and internet-based services all in one place.

HEVC High Efficiency Video Coding is a video compression standard. It has been designed to compress the large amounts of data used in Ultra HD/4K broadcasts.

HD High-Definition.

HDCP High-bandwidth Digital Content Protection is used to prevent unauthorised copy of high-definition material.

HDD Hard Disk Drive.

HDMI High-Definition Multimedia Interface. A high quality digital connection that carries video and audio signals. Widely used with high-definition TV screens.

HDTV High-Definition Television.

HDR High Dynamic Range. A technology found in photography and TV technology. It works to create a single image from a combination of images, using the best parts of each to create a more balanced image, with better shadow and highlight detail.

Hz A measure of how quickly a TV picture is refreshed or scanned. 100Hz is fast, 50Hz is the norm but can cause flicker on some pictures.

IBU International Broadcasting Union.

Television Viewer's Guide

GLOSSARY ABBREVIATIONS

iDTV Integrated Digital TV - a TV with built-in digital terrestrial receiver.

iLINK See Firewire.

Impulsive Interference A term used to describe the interference coming from sources such as electrical switches, central heating timers and car ignition systems. It can cause problems with digital reception.

Interactive Services that enable the viewer to interact with the TV programme to obtain more information or to contact the broadcaster.

IPTV Internet Protocol TV - using the internet to deliver television and video services.

ISP Internet Service Provider - a company which provides internet connections.

ITC Independent Television Commission. Replaced in Dec 2003 by Ofcom.

JPEG Joint Photographic Experts Group. Pronounced jay-peg, JPEG is a commonly used system for compressing (reducing the file size) of digital photographic images.

LCD Liquid Crystal Display - technology used in some computer monitor and TV screens.

LCOS Liquid Crystal On Silicon - home cinema projector technology.

Letterbox The effect, with black bands top and bottom of the screen, when a 4:3 television displays a wider image.

Live Pause A feature found on set-top boxes that record. It enables you to pause a programme and then resume watching, from the point where you left off, while the recorder carries on recording to the end of the programme. Also called Chase Play and TimeSlip.

LLU Local Loop Unbundling is the process whereby telecoms companies and ISPs can put their own equipment into BT's telephone exchanges. They can then offer telephony and broadband services which are not reliant on BT.

LNB Low Noise Block down-converter. Essentially the antenna on a satellite dish that picks up the satellite signal and feeds it via co-ax cable to the satellite receiver.

Loop-through Connections on AV equipment that enable a signal to be passed on to other equipment.

LoveFilm Owned by Amazon, the LoveFilm brand has gone as it has moved into the Amazon Prime service.

Macrovision A video copyright protection system which prevents the copying of material from DVD and videotape.

MATV Master Antenna TV. Aerial system where one aerial feeds several TVs.

MHEG Multimedia and Hypermedia Experts Group international standards including MHEG-5 for interactive multimedia applications.

MHP Multimedia Home Platform, a standard used for interactive services.

MP3 A sound file format for downloading music from the internet and used by small portable music players.

MPEG Motion Picture Experts Group - Various formats, such as MPEG1 and 2. MPEG2 is used for DVDs and is an efficient digital way to store or transmit video information. MPEG4 is a new format that contains a wide variety of compression technologies. It is used with high-definition broadcasts.

Mux or Multiplex A group of channels transmitted in a 'bundle' together.

Netflix An on-demand and internet streaming service providing films and TV programmes.

NOW TV An internet TV service owned by BSkyB.

NVoD Near Video on Demand. A way of broadcasting - found mainly on Sky and cable - where several channels are used to transmit the same programme or film, with staggered start times across the channels.

Ofcom Office of Communications. The UK regulator responsible for TV and radio.

OLED Organic Light Emitting Diode - OLED technology is used in screens for mobile phones and portable digital audio players and increasingly in televisions.

P2P A Peer-to-Peer computer network is a network that relies on the computing power and bandwidth of the participants in the network rather than concentrating it in a relatively few servers.

Pay-Per-View Some broadcasters offer pay-per-view services. You will normally need to have an account set up with the broadcaster and your digibox will need to be connected to a phone line to authorise payment. Usually used for films and sporting events.

PCI Peripheral Component Interconnect, a slot or form of connection found inside a PC - for example can accept video and tuner cards.

PDP Plasma Display Panel.

Pillarboxing The effect when black bars are used to mask the sides of a widescreen TV screen, usually when 4:3 image is being displayed. Also see letterboxing.

Pixel Stands for picture element. Used to describe the individual dots of light or colour that make up a television picture.

Plasma Plasma TV screens are flat and use a gas-discharge system to create the picture.

Pro-Logic Analogue Dolby Surround Sound system.

Progressive Scan The way computers and some TV screens generate their picture.

PSP PlayStation Portable - Sony's portable game and video player.

PSU Power Supply Unit.

PTR Personal Television Recorder. Equipment such as Sky+ or Freeview PVR that records onto a disk.

PVR Personal Video Recorder - see PTR.

QAM Quadrature Amplitude Modulation is a modulation scheme. It is used in Freeview broadcasts in

GLOSSARY ABBREVIATIONS

two variations, 16-QAM and 64-QAM. 16-QAM offers the most robust signal, but 64-QAM is able to carry more channels per multiplex.

RCA/phono An RCA jack, also referred to as a phono connector, is a type of analogue connector commonly used with audio/video equipment.

RCE Regional Coding Enhancement. Regional coding on DVDs.

Red button Method of accessing interactive and enhanced television services by using the red button on a remote control.

Regional coding The system designed to stop DVD discs sold in one part of the world playing on a DVD player from another part of the world.

Return path A channel used to communicate back to the broadcaster for interactive purposes.

REW Rewind.

RF Radio Frequency. RF (co-axial) cables are used to carry signals between aerials, TVs and VCRs.

RGB Red/Green/Blue. RGB Scarts or connections offer medium quality pictures.

S/N Signal-to-Noise ratio.

S-VHS An improved version of VHS recorder - it requires special S-VHS cassettes - now obsolete.

S-Video Higher quality than composite video - a connector used for sending video signals between items of equipment.

S/PDIF Sony/Philips Digital Interface. A system for conveying digital programme and control data.

SACD Super Audio CD. A Sony rival to DVD-Audio. Offers high quality sound - can be played on dedicated machines and some DVD players.

Scart A type of connection and cable system used to carry analogue video and audio signals.

SDTV Standard-definition Television.

SECAM Séquentiel couleur à mémoire. A TV system used in France and Russia. On a UK PAL TV SECAM pictures may appear black and white.

SES Société Européenne des Satellites - owns and controls the Astra group of satellites.

Set-top box or STB. A digital receiver - either satellite, Freeview or cable. A misnomer as they usually sit under the TV! Also called digiboxes.

Sky BSkyB (British Satellite Broadcasting).

Smart A term used to describe TVs, set-top boxes and other equipment that is connected to the internet in order to view online content such as BBC iPlayer.

Smart Card Used to view encrypted programmes.

SMATV Satellite Master Antenna TV. Satellite system where one dish feeds a number of TVs.

Saorview The digital terrestrial TV system used in the Republic of Ireland.

STB See set-top box.

Subcarrier An supplementary satellite signal carrying an audio soundtrack. May carry an additional language commentary or can be used to carry a separate radio station.

Subwoofer Loudspeaker designed to reproduce deep bass sounds.

Switchover The transition from analogue to digital broadcasts. Also called switch-off.

Teletext News and information service carried alongside TV broadcasts.

Terabyte 1000 Gigabytes.

Terrestrial Signals beamed from transmitters on the ground and picked up via an aerial, rather than by a satellite dish.

TETRA Terrestrial Trunked Radio. A digital mobile radio system used by the police and emergency services. Can cause interference with analogue and digital TV.

TFT Thin Film Transistor - a technology found with some LCD-type screens.

THX Home Cinema quality control standard designed by LucasFilm for sound reproduction.

Timeshifting Recording a programme for watching at a later time.

Transponder A satellite's transponders receive signals from earth and re-transmit them back to your dish.

UHF Ultra High Frequency, a TV broadcast band.

Ultra-HD Ultra high definition is also known as Ultra HD, UHD, and UHDTV. It usually refers to devices offering 4K resolution or playback.

UMD Mini-DVD sized disc, used by Sony's portable PlayStation.

USB Universal Serial Bus - connection commonly used to connect devices to computers.

VCR Video Cassette Recorder.

VGA Video Graphics Array. Used to describe systems with 640 x 480 resolution capability.

VHS Video Home System, designed by JVC in the late 70s - it was the most popular format for VCRs.

VoD Video-on-Demand. A term used to describe the ordering and delivery of films or television programmes from a broadcaster. Usually delivered via the internet.

VoIP Voice over Internet Protocol is the routing of voice conversations over the internet.

Widescreen A term used to describe cinema-style wide format pictures - usually 16:9 format.

WiFi A wireless connection used in computing, mobile technology and increasingly with TV equipment. Has a typical range of around 10-20 metres indoors.

XGA Extended Graphics Array, is used to describe displays displaying resolutions up to 1024x768 pixels.

New editions of the guides are published in early December each year. For example, the 2016 editions are published in December 2015.

You may place an order on the website, **www.radioguide.co.uk** or by using the form below.

To order by post please complete the form below

Please send order form and remittance to: **RLG, FREEPOST (SWB40688), PLYMOUTH PL8 1ZZ**

Item:	Qty	Price	Total
Radio Listener's Guide 20 ___ edition		£5.95	
Television Viewer's Guide 20 ___ edition		£5.95	
Mobile Phone Guide 20 ___ edition		£5.95	
How to take Great Photographs		£9.95	
FREE P&P to UK destinations with all orders over £15			
Standard UK delivery (within 7-10 days)			£2.00
		Grand Total	

☐ I enclose a cheque / postal order for £ _____ made payable to **RLG**.

☐ Please charge my credit / debit card.   Visa / Mastercard / Maestro / Solo.

Expiry date _____   Issue no _____ (if applicable)

☐☐☐☐ ☐☐☐☐ ☐☐☐☐ ☐☐☐☐ ☐☐☐☐

Card Number

Name on card _____

Signature _____   Date _____

PLEASE PRINT

Title _____ Initials _____

Surname _____

Address _____

_____

_____

Postcode _____

Tel _____

Email _____

_____

We do not sell our customer information, or make it available to other companies.